ENVIRONMENTAL ENGINEERING LABORATORY MANUAL

Dr. Balamurali S
Mr. Baranitharan B

Copyright © 2016 Balamurali S , Baranitharan B

All rights reserved.

ISBN: **1536911135**
ISBN-13: **978-1536911138**

DEDICATION

This book is dedicated to my father Late Thiru T.Balakrishnan who taught me the values of life.

Baranitharan B

ACKNOWLEGEMENT

First and Foremost I thank the GOD for His abundant blessings. It gives me a great pleasure to avail this opportunity.

I express my deep sense of indebtedness and profound gratitude to **Prof.P. Murugan** Professor, Department of Electronics and communication Engineering, Kalasalingam University, Srivilliputhur, Tamil nadu, India, for his guidance and co-operation in completing the manual work.

I express my sincere thanks to my brother **Mr.B.ANBAZHAGAN** Software Engineer, Japan, for Support and motivate myself.

I am very grateful to faculty of civil Engineering Department, Syed Ammal Engineering College, Ramnad, Tamil nadu, India my friends and my Family for their Moral support.

BARANITHARAN B

Contents

Sl. No.	Name of the test	Page No.
1	Study of sampling and preservation methods and significance of characterization of water and waste water	8
2	Determination of pH	19
3	Determination of Optimum Coagulant Dosage by Jar Test	24
4	Determination of Residual Chlorine	28
5	Determination of hardness	32
6	Determination of chloride	36
7	Determination of Turbidity	40
8	Determination of Available Chlorine in Bleaching Powder	44
9	Determination of Total, Fixed and Volatile Solids	48
10	Determination of Suspended and Dissolved Solids	53
11	Determination of Total Settle able solids	57
12	Determination of Dissolved Oxygen	61
13	Determination of Sulphates	65
14	Determination of Fluorides	69
15	Determination of Ammonical Nitrogen	73
16	Determination of COD	77
17	Determination of Iron	81
18	Determination of Biochemical Oxygen Demand	86
19	Introduction to Bacteriological Analysis	90
20	Definitions	94
21	Objective Questions	103

TEST 1

Study of sampling and preservation methods and significance of characterization of water and waste water

OBJECTIVE

1. To study the sampling and preservation methods in water and waste water characterization.
2. To learn the significance of characterization of water and waste water.

SAMPLING PROGRAMME AND PROCEDURES

The collection of a representative sample is the most important function of an environmentalist. The interpretation of results and recommendation for prevention and corrective treatment are all based on the analysis report. Scrupulous care in the collection of samples is therefore necessary to ensure that the sample is representative of the body of water under examination and to avoid spoilage and accidental contamination of the sample during collection and transport.

METHODS OF SAMPLING

Three types of samples are often collected depending on situations.

a. Grab Samples:

Grab samples are samples collected at a designated place at a particular time. They represent the composition at the time and space. When a source is known to vary in time, as in the case of waste effluents, grab samples are collected at various time intervals and analyzed separately can be of greater value.

b. Composite samples

Composite samples are a mixture of grab samples collected at one sampling point at different times. Individual samples are collected in wide mouth bottles every hour and mixed in volume proportional to the flow. The composite values are useful for observing average values.

c. **Integrated samples**

Integrated samples are a mixture of grab samples collected from different points simultaneously and mixed in equal volumes. Individual samples are collected from both banks of a river and at varying depths to represent available situations.

SAMPLING AND PRESERVATION REQUIREMENTS

1. **Physical and Chemical Requirements:**

For general physical and chemical examination, the sample should be collected in a chemically clean bottle made of good quality glass fitted with a ground glass stopper or a chemically inert polyethylene container. The volume of sample to be collected would depend on the selection of tests; however, for general examination 3.0 liter sample would be sufficient,

The following precautions must be taken while collecting the sample

i) The sampling location is representative of the water body
ii) The place is devoid of floating material
iii) Where ever possible the sample should be collected 15cm, below the surface or as the situation warrants
iv) No physical activity is permitted upstream of sampling point

Shorter the time between collection and examination, the reliable will be the analytical results. For certain constituents and physical values, immediate analysis in the field is required, because, the composition of water may change before it arrives at the laboratory.

The maximum limits of storage are:
Unpolluted water : 72 hours.
Slightly polluted : 48 hours.
Grossly polluted : 12hours.

Some determinations are more likely to be affected by storage than others. Temperature may change, pH may change significantly, and dissolved gases may be evolved and lost (O_2, CO_2, and H_2S).

Frequency of sampling

Frequency depends on objectives. Yet, collection of samples of both raw and treated waters should be carried out as frequently as possible and at least once in every three months. Some waters undergo more pronounced seasonal variation and therefore require more frequent testing. Samples from treatment units should be collected and analyses frequently, at least one from each unit daily.

2. Bacteriological requirements:

The samples for bacteriological examination are collected in sterilized, neutral glass, glass-stoppered 8oz, and 300 ml bottles. The stopper and the neck should be protected by paper or parchment cover. If the sample is likely to contain traces of residual chlorine, an amount equal to 3.0 mg of sodium thiosulfate ($Na_2s_2O_3$, $5H_2O$) to neutralize chlorine is added to the bottle before sterilization. The sterilization is done at 15 psi ($121°C$) for 20-30 minutes in an autoclave.

The sterilized sample bottle should be kept unopened until the time of collection. The stopper should be removed with care to eliminate chances of spoiling and contamination and should never the rinsed. After filling, the stopper should be replaced immediately. The place of collection should be predetermined and procedure of collection conditioned depending on the source.

The standard procedure in sampling from a water faucet or tap is as follows:

a) Flame the tap briefly to kill clinging bacteria. This can be done with a piece of burning paper.
b) Turn on the water and allow it to run for 1 min.
c) Remove the stopper from the bottle, being careful not to touch the inner portions o the stopper or bottle neck.
d) Fill bottle carefully, allowing no water to enter that has come in contact with hands. It is sometimes necessary to collect a sample from a reservoir or basin. If the water can be reached, remove the stopper, plunge the bottle below the surface and move the bottle while it is filling, so that no water will enter that has been in contact with hand. If the water is out of reach, as in a dug well, the bottle can be lowered with a cord.

The sample after collection should be examined immediately, Preferably within one hour. If the conditions do not permit immediate

Examination, the sample should be stored at low temperatures. This period should in case be more than 24 hours. If storage or transportation is necessary, they should be got at a temperature between 0°C and 10°C.

Frequency of sampling:

The frequency of sampling should be fixed depending on the magnitude of the problem involved. The number of samples to be examined from drinking water supply distribution system is normally decided on the basis of population served as given in the tabulation:

Population	Treated/untreated water entering distribution system	
	Max. interval between successive sampling	Max. No. of samples to be Examined.
Up to 20,000	1 month	One sample for every 5000 population
20,001 to 50000	15 days	
50,001 to 100,000	4 days	
More than 100,000	1 day	One sample for every 10,000 population

The raw water should be examined as frequently as the situation demands. The frequency is also determined based on objectives of study.

3. Biological Requirements:

In general the samples for biological examination are collected in wide mouth, clean glass bottles of 2.0 liter capacity. They are never filled completely. This method is employed when total microscopic count is the aim. In some specific cases the concentrate of a sample may be collected through plankton nets made of bolting silk cloth, or the sample filtered through Sedge wick Rafter funnels.

In general the sample must be examined microscopically within one hour of collection. If the facilities do not permit an immediate examination, it should be preserved after collection by addition of 2 ml neutralized (pH 7.0) formalin to each 100 ml of the sample. There is no practice about the frequency of sampling but the examination should be made regularly, or else as the situation demands. Benthos study is

Complex, Collection through cages placed at proper preselected sites for a defined period of time is recommended.

PRESERVATION METHODS FOR CHARACTERIZATION OF SAMPLES

Determination	Containers	Minimum sample size (ml)	Preservation	Max.storage recommended
pH	P,G	-	Analyze immediately preferably in field	30 minutes
Solids	P,G	300	Refrigerate	7 days
Sulfates	P,G	100	Refrigerate	28 days
D.O	G,BOD bottle	300	Titration may be delayed after fixation (1 ml Alk.KI and 1 ml $MnSO_4$) And acidification.	8 hours
Turbidity	P,G	-	Analyze same day, store in dark, refrigerate	24 hours
Hardness	P,G	100	Add HNO_3 to pH<2	6 months
Fluoride	P	300	None	28 days
COD	P,G	100	Analyze as soon as possible or add H_2SO_4 to pH<2	7 days
B.O.D	P,G	1000	Refrigerate	6 hours
Chlorine residuals	P,G	500	Analyze immediately	30 minutes
Ammoniacal Nitrogen	P,G	500	Analyze as soon as possible or add H_2SO_4, refrigerate	6 months

Note: Refrigerate - Storage at 4°C; P = plastic (polyethylene or equivalent), G = Glass, neutral.

SIGNIFICANCE OF CHARACTERISATION OF VARIOUS PARAMETERS

Natural waters are never completely pure. During their precipitation and passage over or through ground they require a wide variety of dissolved and suspended impurities. The concentrations of these impurities are seldom large in ordinary chemical sense but they modify the chemical behavior of water or its usefulness.

Hardness

The study of hardness is important from the point of view of industrial utilization of water specially in boilers, where scales are formed. Hardness in municipal supplies increases the consumption of soap, fuel, tea leaves etc. in the household an renders it unsuitable for use in air-conditioning.

Turbidity

It is a measure of degree of opaqueness of water and interference presented by suspended matter to the passage of light. The turbidity is due to clay, silt, finely divided organic matter and microscopic organisms. Turbidity tests are important from aesthetic consideration and from the point of economics of treatment. The most important health significance of turbidity is that may harbor pathogenic organisms.

Residue or solid matter

The test for residue is of very great importance in sewage treatment processes to indicate the physical state of the principal constituents. The ratio of the weight of suspended solids to turbidity often referred as coefficient of fineness. The solids present in dissolved form are related to the electrical conductivity. The fixed solids indicate the mineral level while volatile solids are related to organic matter.

Some of these impurities are toxic, some may affect health, and some affect the portability while others indicate pollution. A list of such impurities is given below:

Toxic substances	Max. allowable limit (W.H.O Standards) mg/L
Lead	0.05
Arsenic	0.05
Selenium	0.01
Chromium	0.05
Cyanide	0.2
Cadmium	0.01
Components hazardous to health	
Fluorides	1.5
Nitrates as NO_3	45
Compounds affecting the portability	
TDS	1500
Iron	5.0
Manganese	5.0
Copper	1.5
Zinc	1.5
Magnesium plus –sodium sulfate	1000
Surfactants(ABS)	0.5
Chemical indicators of pollution	
BOD	60
COD	10.0
Total Nitrogen exclusive of NO_3	1.0
Ammonia Cal Nitrogen as NH_3	0.5
Carbon chloroform extract(CCE)	0.5
Oil and Grease	1.0
D.O	40% saturation

pH

Determinations of pH, alkalinity and its forms, along with acidity are of interest in coagulation, softening and corrosion control. The balance of positive hydrogen ions (H^+) and negative hydroxide ions (OH^-) in water determines how acidic or basic the water is. Notice the ' + ' and ' − ' in the chemical symbols above. They indicate that these chemical forms are 'ions' — they have a positive or negative electrical charge. This means the molecule in question is either missing an electron or has an extra electron. Since electrons have a negative charge, an extra one in the OH molecule makes it OH^-, and a missing one in the H molecule gives it a "missing-minus" charge — in other words, positive — and makes it H^+. When analysts measure pH, they are determining the balance between these ions.

The pH scale ranges from 0 (high concentration of positive hydrogen ions, strongly acidic) to 14 (high concentration of negative hydroxide ions, strongly basic). In pure water, the concentration of positive hydrogen ions is in equilibrium with the concentration of negative hydroxide ions, and the pH measures exactly 7.

pH scale

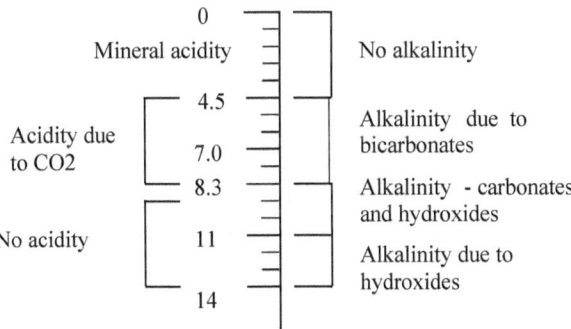

Chloride

Concentration of chlorides in municipal sewage is often significantly (15-50 mg/L) higher than those in its water supply. For this reason, a change in its concentration may be indicative of sewage pollution, in waters of low chloride concentration. Chlorides occur in an all-natural waters in widely varying amounts. Mountain streams are normally low in chloride values. Chloride gain access to water either because of excellent solvent properties or through human excreta or industrial pollutants. Chlorides were for several years used as an indicator of pollution by municipal wastes in rivers, streams, wells and lakes.

Dissolved Oxygen

In raw water and domestic wastes, dissolved oxygen is a factor which determines whether the biological processes undergoing a change are aerobic or anaerobic. It is very desirable that aerobic conditions are maintained. It is a Single test which will immediately indicate the sanitary status of a stream. Low values of dissolved oxygen adversely affect the potability of water and may cause fish kill.

Organic matter

The tests of organic matter indicate type and extent of pollution, which has its origin in plant or animal matter. Tests are mostly restricted to the study of nitrogen in various forms and oxygen requirements in biodegradation of putrescible carbonaceous organic matter (BOD).A measure of the demand is also indicated in terms of demand through strong chemical oxidants (COD).

BOD

The BOD is the amount of oxygen required by bacteria while stabilizing decomposable organic matter under aerobic conditions. Polluted water does not contain sufficient oxygen in solution to maintain aerobic condition during decomposition. The quantity of oxygen required for complete stabilization is taken as a measure of its organic content

COD

The COD test is based on the concept that a large majority of organic compounds can be completely oxidized by the action of strong oxidizing agents in acidic medium. The quantity of oxygen required is proportional to organic matter, regardless of the biological assimilability of the substance.

Nitrogen

Nitrogen is estimated as organic nitrogen, ammoniac nitrogen, nitrite nitrogen and nitrate nitrogen throw light on the pollution history of the carrying water.

Ammonia

Ammonia is deadly. It is primarily created through the fish's gills, kidneys and intestinal waste. Decaying uneaten food, plant material and leaves also contribute to ammonia accumulations. Ammonia is reportedly the leading cause of fish stress, breaking down his immunity system and leading to bacterial disease. Measurable to high levels of ammonia is common in new ponds (and aquariums), over-stocked ponds and established ponds from heavy feeding in the spring prior to biological bacteria growth or from inadequate filtration. *The only acceptable reading from an ammonia test is "0".*

Nitrites

Nitrites create the first step in the nitrifying cycle and are second to ammonia in its toxicity to fish. A Nitrosomonous bacterium converts ammonia to nitrites. Nitrite readings are normally seen as the nitrifying cycle is developing and can be present without an ammonia reading. *The only acceptable reading from a nitrite test is "0".*

Nitrates

Nitrates are the last step in the nitrifying cycle. Nitrobacteria bacteria convert the nitrites to nitrates. As a rule, nitrates are nontoxic to fish. High nitrate levels contribute to algae bloom. Basically, nitrates are fertilizers. *Acceptable test readings are 200 to 300.*

Bacteriological tests

The routine bacteriological tests are aimed at enumerating the members of coliform group, which are considered indicators of pollution. The natural habitat of these bacteria is the intestinal tract of man and other warm blooded animals. They are present wherever the pathogens are present and by their absence exclude the probability of the presence of pathogens. They share the fate of the most significant pathogenic enteric bacteria outside the human and animal body both in the rate of death and in the rate of removal when water is purified.

Another test of bacteria is aimed at detecting chemo-synthetic heterotrophic heterogeneous group developing under conditions of cultivation and is referred as Total Plate Count. This test is not differential and indicates a total picture of bacteria associated with organic matter.

Biological Examination

The biological examination (microscopic) provides useful information for the control of water quality and treatment. It serves for one or several of the following purposes:

i) To explain the cause of color or and odor in water.
ii) To aid in the interpretation of various chemical analysis reports.
iii) Permitting identification of specific water when it is mixed with another.
iv) To explain clogging of pipes/screens/filters.
v) Rapidly detect organic pollution and contamination with toxic substances.
vi) To indicate the progress of self-purification streams.

RESULT

Thus the sampling and preservation methods in water and waste water characterization and to learn the significance of characterization of water and waste water are studied.

TEST 2

Determination of pH

OBJECTIVE

To determine the pH value of the given sample by electrometric method.

APPARATUS REQUIRED

pH meter with combined electrode, beakers.

CHEMICALS REQUIRED

Buffer tablet of pH values 4 and 9.2.

REAGENTS PREPARATION

Buffer solution of pH value 4

Buffer tablet of pH value 4 is dissolved in 100 ml of distilled water. This solution should preferably be stored in a plastic bottle in cool place.

Buffer solution of pH value 9.2

Buffer tablet of pH value 9.2 is dissolved in 100 ml of distilled water. This solution should preferably be stored in a plastic bottle in cool place.

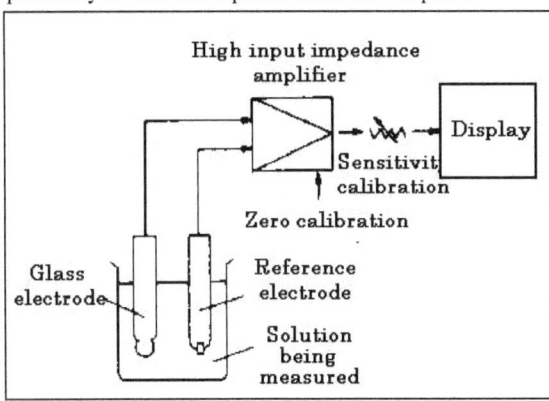

PROCEDURE

Electrometric method:

1. Wash the combined electrode of pH meter with distilled water and clean the same with distilled water.
2. Dip the combined electrode in the buffer solution of pH value 4.
3. Adjust the temperature by the adjustment knob to an ambient (room) temperature.
4. If the instrument shows the reading as 4 then it is in order if not, adjust the reading to 4.0 by calibration adjustment knob.

5. Wash the electrode of pH meter with distilled water and clean the same with distilled water and dip it to the buffer solution of pH value 9.2.
6. Note the reading if the instrument shows the reading as 9.2 then it is in order otherwise uses the calibration adjustment knob and bring the reading to 9.2.
7. Repeat the above procedure until the meter shows reading as 4 when electrode is dip in buffer solution of pH 4 and shows reading as 9.2 when electrode is dip in buffer solution of pH value 9.2.
8. Now the instrument is calibrated.
9. After cleaning the electrode dip in the sample for which pH value is to be found out.
10. Directly record the reading from the meter without doing any adjustments.

ENVIRONMENTAL SIGNIFICANCE

pH (6.5 to 8.5) has no direct effect on health however a lower value below 4 will produce sour taste and higher value above 8.5 a bitter taste. Higher values of pH have scale formation in water heating operators and also reduce the germicidal potential of chlorine. High pH induces the formation of trihalomethanes which are causing cancer in human beings.

pH below 6.5 starts corrosion in pipes, thereby releasing toxic metals such as zinc, lead, cadmium & copper etc., According to BIS water for domestic consumption should have pH between 6.5 to 8.5.

Application of pH data in environmental engineering practice

1. Determination of pH is one of the important objectives in biological treatment, if the pH goes below 5 due to excess accumulation of acids, the process is severely affected. Shifting of pH beyond 5 to 10 upsets the aerobic treatment of the waste waters. In these circumstances, the pH can be adjusted by addition of suitable acid or alkali to optimize the treatment of waste water.
2. Its range is of immense value for any chemical reaction. A chemical value shall be highly effective at particular pH. Chemical coagulation; disinfection, water softening and corrosion control are governed by pH adjustment.
3. Dewatering of sludge's, oxidation of cyanides and reduction of hexa covalent chromium in to trivalent chromium also need a favorable range 4. It is used in
the calculation of carbonate, bicarbonate, CO2 calculation, stability index and acid-base equilibrium.

OBSERVATION

Source of sample :
Date of collection :
Time :

TABULATION

Sl. No.	Sample	Calibration with	Temperature (^{0}C)	pH value	Remark

RESULT

The pH value of the given sample by electrometric method is_____.

TEST 3

Determination of Optimum Coagulant Dosage by Jar Test

OBJECTIVE

To determine the optimum dosage of coagulant required for a given sample of waste water.

APPARATUS REQUIRED

1. Laboratory flocculator with stirring paddles.
2. Glass Jars.
3. Analytical balance.

REAGENTS REQUIRED

Alum solution (made by dissolving 14.28gm of alum in one liter distilled Water).

THEORY

Alum is the name given to the Aluminum sulphate with its chemical formula as $Al_2(SO_4)_3.18H_2O$. The alum when added to raw water, reacts with the bicarbonate alkalinities, which are generally present in raw waters, so as to form a gelatinous precipitate (floc) of aluminum hydroxide. This floc attracts pure suspended matter and colloids present in raw water, thereby growing in size. The floc formation is assisted by slow mixing, called floatation. The flocs finally settle down to the bottom of the tank, for being removed in the sedimentation tank. The above process/technique is known as coagulation.

The chemical equation involved in coagulation is given below:
$$Al_2(SO_4)_3.18H_2O + 3Ca(HCO_3)_2 \longrightarrow 3CaSO_4 + 2Al(OH)_3 \downarrow + 6C_2O \uparrow$$

The amount of coagulant required for coagulation depends on the turbidity of the waste water. The use of optimum amount of coagulant is indicated by the formation of the large feathery flakes. This can be approximately determined in the laboratory by Jar test, which can finally be adjusted with actual results obtained at the treatment plant. The optimum dose of alum is usually found to vary from 5mg/l for relatively clear waters to about 85mg/l for highly turbid waters. The average normal dose is about 17mg/l.

The jar test involves the use of a stirring device shown in Fig. the stirrer consists of six paddles capable of rotation with variable speed between 0 to 100 rpm. In the test, 1 liter water is placed in each of the jars or beakers, which are dosed with different amounts of coagulant (alum). After rapid mixing to disperse the chemicals, the samples are stirred slowly for floc formation; and then allowed to settle under quiescent conditions. For this purpose, the jars are initially mixed at a speed of 60 to 80 rpm for 1 minute, and the stirred at a speed of 30rpm for 15 minutes. After the stirrer is stopped, the nature and settling characteristics of the floc are observed and recorded in quantitative terms, such as poor, fair, good, very good or excellent. A hazy sample indicates poor coagulation, while proper coagulated water contains floc that is well formed with the liquid clear between particles. The lowest alum dosage that provides good turbidity removal during the jar test is considered for the first trial dosage in plat operation, and the final optimum quantity is adjusted by actual observations at the WTP (Water Treatment Plant)

PROCEDURE

1. Fill 1 liter waste water sample in each of the six jars.
2. Attach the sample jars to the stirring device by lifting the paddles in the right upward direction.
3. Add coagulant (Alum) in progressive dosages into the series of the six sample jars.
4. The coagulant dosage can be selected randomly depending on the characteristics of waste water.
5. Mix the sample rapidly for about 10 min with mechanically operated paddles at 180rpm followed by gentle stirring about 10 min. at 30 – 40 rpm.
6. Remove the jars from the stirring device after stirring is completed.
7. Let the sample in the jars stand for 30 min. for settling of flocs.
8. The dose of coagulant versus floc formation is plotted as graph.

9. The dose of coagulant which gives the best floc is the optimum dose of coagulants.

OBSERVATION

Volume of sample : **A ml**
Source of sample :
Date of collection :
Time :
Temperature :

TABULATION

Sl. No.	Jar No.	Amount of coagulant added B (gm)	Floc formation (ml)

CALCULATION

Optimum Dosage = B/A mg/L

RESULT

The optimum coagulant dosage = _____ mg/L.

TEST 4

Determination of Residual Chlorine

OBJECTIVE

To determine the residual chlorine for the given water sample.

APPARATUS REQUIRED

Burette with stand, tiles.
Pipette.
Conical flask.
Beaker.
Glass funnel.
Measuring jar.

PRINCIPLE

Chlorine combines with water to form Hypochlorous and Hydrochloric acids $Cl_2 + H_2O \rightarrow HOCl + H^+ + Cl^-$

In water chlorine, hypochlorous acid and hypo chlorite ions are referred as free chlorine residuals and the chloramines are called combined chlorine residuals.

The chlorine demand of water is the amount of chlorine that must be supplied to leave a desired free combined or total residual after a specified contact period.

The starch iodide test is an age old method for testing the total chlorine residual in a given water sample and is still being used, depending upon the oxidizing power of free and combined chlorine to convert iodide ion to free iodide. This free iodine liberates iodine ions when titrated with sodium thio sulphate as shown below;

$Cl_2 + 2I^-$ ---------- $I_2 + 2Cl^-$ I_2 + Starch --------- Blue colour $I_2 + Na_2S_2O_3$ -------- $Na_2S_4O_6 + 2NaI$ (Or)

$I_2 + 2S_2O_3$ ---------- $S_4O_6 + 2I^-$

REAGENTS REQUIRED

Chlorine water 1 gm/L.
Acetic Acid.
Standard N/40 Sodium Thiosulphate solution.
Starch indicator.

Reagents Preparation:
Starch Indicator:

Weigh 1 gm of starch and make it into a paste with 10 ml of hot water and dilute it to 100 ml.

Standard N/40 Sodium Thiosulphate solution:

Dissolve 1.575 gm of $Na_2S_2O_3$ in distilled water and make up to 1 liter.

PROCEDURE

1. Take 25 ml of given water sample in a conical flask.
2. Add a small crystal of KI and distilled water to the above flask containing water sample, to make 100ml.
3. Add about 0.5 ml conc.HCl or about 10ml acetic acid to act as buffer to reduce the pH to a low value between 3.5 to 4.2 to avoid conversion of Cl_2 into HOCl and OCl^-
4. Titrate the above yellow coloured iodine solution against standard Sodium thio sulphate till yellow becomes light or faintly yellow.
5. Add 1 ml of soluble starch solution (end point indicator) to change the colour from light yellow to blue in the conical flask.
6. Continue the titration with standard sodium thio sulphate till the blue colour just disappears.
7. Note down the total amount of titrant used in the entire titration. Let it be 'X'ml.
8. Repeat the titration with distilled water as sample and determine the amount of sodium thiosulphate let it be 'Y' ml.

ENVIRONMENTAL SIGNIFICANCE

The residual chlorine is the measure of chlorine left in water after the required contact period, which will ensure complete killing of bacteria and oxidation of the organic matter. Usually a free chlorine residue of 0.2 to 0.3 mg/L after a contact period of 10-20 minutes is considered to be sufficient and satisfactory to take care of the future contamination of water to be supplied through the distribution system.

OBSERVATION

Source of sample
Date of collection
Time
Temperature

TABULATION

Sl.No	Quantity of water Sample used in ml	Test on water sample ml(Burette Reading)			Test on distilled water ml(Burette Reading)			Chlorine residuals mg/L
		Initial	Final	$Na_2S_2O_3$ used (X ml)	Initial	Final	$Na_2S_2O_3$ used (Y ml)	

CALCULATION

Residual Chlorine in original water sample

$$= \frac{(X-Y) * 1000 * 0.8895}{\text{ml of water sample}}$$

= _____ mg/L.

RESULT

Residual Chlorine residuals for given water sample by Starch Iodide Test is -------mg/L.

TEST 5

Determination of hardness

OBJECTIVE

To determine the total hardness present in the given sample.

THEORY

Hardness in water is that characteristic which prevents the formation of sufficient foam, when such hard waters are mixed with soap. The hardness is usually caused by the presence of calcium and magnesium salts present in water, which form scum by reaction with soap.

Hard waters are undesirable because they may lead to greater soap consumption, scaling of boilers, erosion and incrustation of pipes and making food tasteless etc.

The bicarbonates and carbonates of the divalent metallic ions, principally of Ca^{++} and Mg^{++}, cause carbonate hardness, while their sulphates, chlorides and nitrates cause non-carbonate hardness.

Hardness of water is usually measured in mg/l of $CaCO_3$, and depending upon their hardness, the waters are usually classified as follows:

Sl. No.	Category of hardness	Hardness in mg/L
1	Soft	0 – 75
2	Moderately hard	75 – 150
3	Hard	150 – 300
4	Very hard	>300

The total hardness of a given sample of water can be easily computed by titration method, by using the E.D.T.A (Ethylene-Diamine-Tetra-Acidic acid) solution, or its sodium salt [ie. Disodium ethylene diamine tetra acetate.

APPARATUS REQUIRED

Burette, Pipette, Conical flask, measuring jar.

CHEMICALS REQUIRED

Eriochrome Black-T (EBT) indicator, Ammonium chloride, Ammonium solution, EDTA.

REAGENTS PREPARATION

EBT indicator
Dissolve 0.2 gm of pure solid in 15 ml of distilled water.

Standard EDTA Titrant
Take 3.7gm of the dissolve and add to get distilled water to make 500ml. 1 ml of exactly 0.02 N EDTA=1mg of $CaCO_3$.

Ammonia Buffer solution
Dissolve 7gm of ammonium chloride (NH4Cl) in 56.8ml concentrate ammonia solution and dilute to 100 ml.

PROCEDURE

1. Take 20 ml of the sample in a conical flask.
2. Add 2ml of ammonia buffer to the flask.
3. Add 5 – 6 drops of EBT indicator to the flask wine red colour will be developed.
4. Titrate it with standard EDTA solution which is filled in the burette till the colour changes from wine red to blue.
5. Repeat steps1 to 4 for different samples with varying hardness and also for distilled water (blank).
6. For determination of non-carbonate hardness the sample is to be boiled for 30 minutes. The procedure is the same as above.

SANITARY SIGNIFICANCE

Hard water has adverse action with soap since it allows less formation of leather. If hard water is used in boilers, scaling problem occurs leading to the bursting of boilers, It makes food tasteless. It affects the working of dyeing process. It is also precipitate protein of meat and make tasteless.

Application of Hardness data in Environmental Engineering Practice

1. Hardness of water is important in determining the suitability of water for domestic and industrial uses.
2. The relative amounts of calcium and magnesium hardness, carbonate and non-carbonate hardness present in water are the factors while determining the most economical type of softening process.
3. Determination of hardness serve as a basis for routine control of softening process.

OBSERVATION

Source of sample

Date of collection

Time

Temperature

TABULATION

Sl.No	Vol. of water sample (ml)	Burette Reading(ml)		EDTA consumed (ml)	Hardness (mg/L)
		Initial	Final		

CALCULATION

$$\text{Hardness in mg/l of } CaCO_3 = \frac{\text{ml of EDTA used}}{\text{ml of water sample}} \times 1000$$

RESULT

Total Hardness in mg of $CaCO_3$ = _____ mg/L.

TEST 6

Determination of chloride

OBJECTIVE

To determine the amount of chloride present in the given sample.

THEORY

Chlorides are found to occur in all the natural waters, and their quantity may vary widely, depending upon various factors. For example, upstream hilly reaches of rivers may have lesser chlorides while the downstream reaches may have more chlorides, since the solvent action of water may dissolve a lot of chloride from the top soil, over which the water travels during its journey. The sewage and industrial wastewaters finding their way into the river may also increase the chloride content of the raw water.

Chlorides in given water can be easily measured by volumetric procedures employing internal indicators. Mohr's method, employing silver nitrate solution as a titrant and potassium chromate as the indicator is satisfactory.

APPARATUS REQUIRED

Burette with stand, pipette, conical flask, measuring jar etc.,

CHEMICALS REQUIRED

Sodium Chloride, Silver nitrate, Potassium Chromate.

REAGENTS PREPARATION

Silver Nitrate Solution

Dissolve 1.2gm of silver nitrate in distilled water and make up to 250ml.

Sodium chloride Solution (0.028N)
Dissolve 0.164gm of sodium chloride in distilled water and make up to 1000ml

Potassium Chromate Solution (K_2CrO_4)

Dissolve 2gm of potassium chromate in 20ml of distilled water.

PROCEDURE

Standardization of Silver Nitrate Solution

1. Pipette 20 ml of sodium chloride solution in to the conical flask.
2. Add one or two drops of potassium chromate solution.
3. Titrate against Silver Nitrate solution until the appearance of reddish brown colour.
4. Repeat the titration for concordant values.

Silver Nitrate vs. Sample

1. Pipette 20 ml of sample in the conical flask.
2. Add one or two drops of potassium chromate solution
3. Titrate against silver Nitrate solution until the appearance of reddish brown colour.
4. Repeat the titration for concordant values.

ENVIRONMENTAL SIGNIFICANCE OF CHLORIDES

Chloride associated with sodium exerts salty taste, when its concentration is more than 250 mg/l.There is no known evidence that chloride constitute any human health hazard. For this reason, chlorides are generally limited to 250 mg/L in supplies intended for public use. In many areas of world where water supplies are scarce, sources containing as much as 2000mg/L are used for domestic purposes without the development of adverse effect once the human system becomes adapted to the water.

It can also corrode concrete by extracting calcium in the form of calcide.Magnesium chloride in water generates hydrochloric acid after heating which is also highly corrosive and create problems in boilers.

Application of chlorides data in environmental engineering practice

1. Chlorides determination in natural waters is useful in the selection of water supplies for human use.
2. Chlorides determination is used to determine the type of desalting operators to be used.

3. The chloride determination is used to control pumping of ground water from locations where intrusion of sea water is a problem.
4. Chlorides interfere in the determination of COD a correction must be made on the basis of the amount of chloride present.

OBSERVATION

Source of sample
Date of collection
Time
Temperature

TABULATION

Silver Nitrate Vs. sodium chloride

Sl. No.	Vol. of sodium chloride (ml)	Burette reading (ml) Initial	Burette reading (ml) Final	AgNo3 consumed (ml)

Silver Nitrate Vs. water sample

Sl. No.	Vol. of water sample (ml)	Burette reading (ml) Initial	Burette reading (ml) Final	AgNo3 consumed (ml)	Chlorides (mg/l)

CALCULATION

Normality of $AgNo_3$ (N) = volume of sodium chloride * Normality of NaCl/Vol of $AgNo_3$

Chloride (mg/l) = Volume of $AgNo_3$ x 35.45 x N1 x 1000/ volume of water sample/ volume of water sample

RESULT

Amount of chloride present in the given sample=_____ mg/L

TEST 7

Determination of Turbidity

OBJECTIVE

To find out the turbidity of given sample.

PRINCIPLE

When light in passed through a sample having Suspended particles, some of the light in Scattered by the particles. The scattering to the light is generally proportional to the turbidity. The turbidity of sample is thus measured from the amount of light scattered by the sample, taking a reference with standard turbidity suspension.

APPARATUS REQUIRED

Nephelometric turbid meter, Sample tubes.

REAGENTS PREPARATION

1. Dissolve 1.0gm Hydrazine sulphate and dilute to 100ml
2. Dissolve 10gm Hexamethylene Tetra mine and dilute in 100ml
3. 5ml of each of the above solution (1 and 2) in a 100ml volumetric flask and allow to stand for 24 hrs at 25±3°C and dilute to 1000ml. This solution has a turbidity of 40NTU.

PROCEDURE

1. The Nephelometric turbid meter in switched on and waited for few minutes till it warms up.
2. The instrument is set up with a 40NTU standard suspension
3. The sample is thoroughly shake and kept it for sometimes so the air bubbles are eliminated

4. The sample is taken in Nephelometer sample tube and the sample is put in Sample chamber and the reading is noted directly.
5. If turbidity is too high, then the sample is diluted with turbid free water and again the turbidity is read.

ENVIRONMENTAL SIGNIFICANCE

Turbidity is objectionable because of
a. Aesthetic considerations.
b. Engineering considerations.

When turbid water in a small, transport container, such as drinking glass is help up to the light, an aesthetically displeasing opaqueness or 'milky coloration is apparent.

The colloidal material which exerts turbidity provides adsorption sites for chemicals that may be harmful or cause undesirable tastes and odours & for biological organism that may be harmful. Disinfections of turbid water is difficult because of the adsorptive characteristics of some colloids and because the solids may partially shield organisms from disinfectant.

In natural water bodies, turbidity may impart a brown or other colour to water and may interfere with light penetration and photosynthetic reaction in streams and lakes.

Turbidity increases the load on slow sand filters. The filter may go out of operation, if excess turbidity exists.

Application of Turbidity Data in Environmental Engineering Practice:

Turbidity measurements of particular importance in the field of water supply. They have limited use in the field of domestic and Industrial waste treatment.

1. Knowledge of the turbidity variation in raw water supplies along with other information is useful to determine whether a supply repairs Special treatment by chemical coagulation and filtration before it may be used for a public water supply.
2. Turbidity measurements are used to determine the effectiveness of the treatment produced with different chemicals and the dosages needed.

3. Turbidity measurements help to gauge the amount of chemicals needed from day-today in the operation of water treatment works.
4. Measurement of turbidity is settled water prior to filtration is useful in controlling chemical dosages so as to parent excessive loading or rapid sand filters.
5. Turbidity measurements of the filtered water are needed to check o faculty filter operation.
6. Turbidity measurements are useful to determine the optimum dosage of coagulants to treat the domestic and Industrial wastes.
7. Turbidity determination is used to evaluate the performance of waste treatment plants.

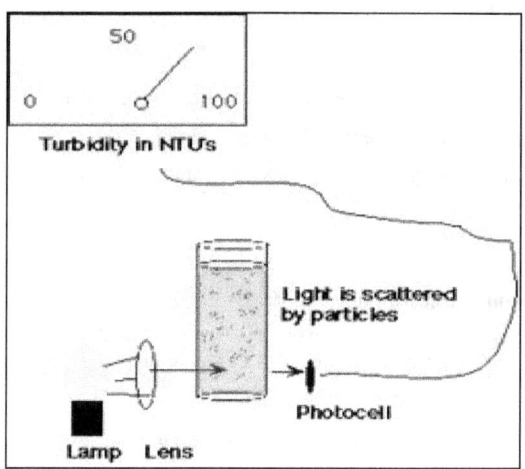

OBSERVATION:

Source of sample:

Date of collection:

Time   :

Temperature   :

TABULATION:

Sl. No	Sample details	Turbidity(NTU)

$NTU = A*(B+C)/C$

Where

    A = NTU of diluted sample.

    B = Volume of dilution water.

    C = Sample volume taken for dilution, ml.

RESULT

    The Turbidity of

    (i)   Tap water ………..

    (ii)   Synthetic sample…………

TEST 8

Determination of Available Chlorine in Bleaching Powder

OBJECTIVE

To determine the available chlorine percentage in a given sample of bleaching powder.

THEORY

In order to work out the disinfecting power of bleaching powder and its requirement for treating given water having a particular chlorine demand, we have to work out the chlorine content in the given bleaching powder.

REAGENTS PREPARATION

Starch Indicator:

Weigh 1 gm of starch and make it into a paste with 10 ml of hot water and dilute it to 100 ml.

Standard N/40 Sodium Thiosulphate solution:

Dissolve 1.575 gm of $Na_2S_2O_3$ in distilled water and make up to 1 liter.

PROCEDURE

1. Take 0.7 gm of bleaching powder in to a beaker.
2. Thoroughly mix the powder with distilled water in the beaker and pour the solution with several rinsing into a 200ml volumetric flask.
3. Fill the 200ml flask with distilled water up to the mark, as to make the chlorine solution equal to 200ml. Mix thoroughly.
4. Dissolve about 2 gm of potassium iodide and 2ml of glacial acetic acid in 2 ml of distilled water in a conical flask.
5. Pipette out 25ml of the chlorine solution from the volumetric flask and add it into the above flask containing iodide acetic acid mix.
6. Add a few drops of starch indicator to the conical flask to develop blue colour.

Date of collection:

Time   :

Temperature   :

Mass of powder used in preparing chlorine solution in beaker and then into volumetric flask=0.7 gm

Vol.of chlorine in volumetric flask=200ml

Vol.of chlorine solution pipetted out in to the conical flask for titration=25ml

TABULATION

Sl.No.	Burette Reading(ml)		Vol. of $Na_2S_2O_3$ consumed X (ml)	% of chlorine in powder sample	End point
	Initial	Final			
					Disappearance of blue colour

CALCULATION

Chlorine in mg/L in chlorine solution = ml of $Na_2S_2O_3$ consumed

Percentage chlorine in Sample= mg/l of chlorine / mg/l of powder * 100

7. Titrate this blue solution in the conical flask against the sodium thiosulphate solution, till the blue colour in the flask just disappears.
8. Note down the ml of sodium thiosulphate solution used in the above titration. It may be averaged out by performing the above test twice or thrice. Let this value be X ml.

RESULT

The percentage chlorine content of the given bleaching powder works out to be =
------ %

TEST 9

Determination of Total, Fixed and Volatile Solids

OBJECTIVE

To determine the amount of total, fixed and volatile solids present in the given sample.

APPARATUS REQUIRED

Crucible, Chemical balance, hot air oven, muffle furnace, desiccator

PROCEDURE

1. Take the empty crucible. Clean it thoroughly and make it perfectly dry. Take the weight of empty crucible.
2. Add to the crucible 20ml of liquid sample.
3. Heat the crucible in water bath at $100^\circ C$ till the entire liquid in a crucible evaporates and dry residue remains at the bottom then place the crucible in oven at $103^\circ C$ for I hour.
4. Take the weight of the crucible with residue after cooling it in a desiccator for 20 minutes. Let us weight be W_2 gm.
5. Take the sample crucible and keep it in a muffle furnace at a temperature of $650^\circ C$ for 30min.
6. The volatile and organic matter in the solids evaporated and the crucible contains only fixed solids.
7. Cool the crucible in a desiccator and weight it with the fixed solids residue. Let the weight be W_3 gm.

ENVIRONMENTAL ENGINEERING SIGNIFICANCE

The water which contains of high volatile solids is not suitable for drinking purposes. The result of high volatile solids indicates that the water may have been pollutes by domestic waste or other organic waste. In general, ground water is free from volatile solids unless they have been polluted by waste seepages. But, well
water may have high volatile solids due to lack of proper protection around well to prevent seepage of used water. Surface water may also have high volatile solids due to disposal of domestic and other wastes.

Application in Environmental Engineering Practice:
1. Volatile solids test is normally applied to sludge's.
2. It is indispensable in the design operation of sludge digestor, vacuum filter and infiltration plants,
3. Before the development of C.O.D test it is used to find the strength of industrial and domestic waste water.
4. It is helpful in accessing the amount biologic ally inert organic matter, such as lignin in the case of wood pulping waste liquor.

OBSERVATION

Source of sample

Date of collection

Time

Temperature

TABULATION

Sl. No.	Vol. of given sample ml	Weight of empty crucible W_1 (gm)	Weight after one heating W_2 (gm)	Weight after two heating W_3 (gm)

Total solids in mg/L = W2-W1/ Vol of sample

RESULT

Total Solids = --------------

Fixed solids = --------------

Volatile solids = --------------

TEST **10**

Determination of Suspended and Dissolved Solids

OBJECTIVE

To determine the amount of dissolved solids present in the given sample.

APPARATUS REQUIRED

Crucible, oven, desiccators, chemical balance.

PROCEDURE

1. Take a known quantity of liquid sample in a crucible of known weight.
2. The sample is filtered through watt man paper number 44. The dissolved solids go in solution through the filter paper.
3. Take a known quantity of filtered solution in a crucible of known weight and dry it to a temperature of $103^{\circ}C$ to $105^{\circ}C$.
4. Cool the crucible in a desiccators and weigh it let the weight be W_2 gm.

ENVIRONMENTAL SIGNIFICANCE

Water with high dissolved solids generally is of inferior portability and makes it use an unfavorable physiological reaction in a transit consumer. Suspended Solids containing much organic matter make as purification and consequently this may be dividing of dissolved oxygen loading to destruction of plant and human life.

Application

Dissolved solids determination gives an idea about the formation of scales cause of foaming in boilers, acceleration of corrosion and interference with the colour and taste of many finishes products.

The suspended solid determination is particularly useful in the analysis of sewage and other waste water. It is used to evaluate the strength of waste water and to determine the efficiency of treatment units.

OBSERVATION

Source of sample:
Date of collection:
Time   :
Temperature   :

TABULATION:

Sl.No.	Vol. of given sample after filtration (ml)	Weight of empty crucible W_1 (gm)	Weight after one heating W_2 (gm)

Dissolved solids in mg/l = W_2-W_1/Vol of sample * 10^6

Suspended solids = Total solids − Dissolved solids

RESULT

Solids present in the given sample

Dissolved solids   =   _____

Suspended solids   =   _____

TEST 11

Determination of Total Settle able solids

OBJECTIVE

To find out the total settle able solids of the given sample.

APPARATUS REQUIRED

Imhoff cone, holding device

PROCEDURE

1. The imhoff cone is gently filled with the thoroughly well mixed sample usually 1 liter and allowed it to settle.
2. After 45 minutes, the cone is rotated between hands to ensure that all solids adhering to the sides are loosened.
3. The solids are allowed to settle for 15 min more to make up for a total period of 1 hour.
4. The volume of the sludge which has settled in the open is noted.
5. The results are expressed in ml settle able solids per liter of sample per hour.

PRECAUTIONS

1. The imhoff cones must be cleaned with a strong soap and hot water using a brush.
2. The cone is wetter before use, which helps in preventing adherence of the solids to the sides.
3. The method is subjected to considerable in accuracy if the solids contain large fragments.
4. The determination of total settle able solids should be carried out soon after sampling in order to avoid errors through flocculation.

OBSERVATION

Source of sample:

Date of collection:

Time   :

Temperature   :

TABULATION

Sl.No.	Sample details	Vol. of given sample (ml)	Total settle able solids (ml/l/hr.)

Application:

1. The settle able solids determination is used extensively in the analysis of industrial waste to determine the need for and design of plain settling tank in plants employing biological treatment process.

2. It is also widely used in waste water treatment plant operation to determine the efficiency of sedimentation tanks.

RESULT

The total settle able solids is = _____ ml/l/hr.

TEST   12

Determination of Dissolved Oxygen

OBJECTIVE

To determine the amount of Dissolved Oxygen present in the given sample.

APPARATUS REQUIRED

Burette with stand, pipette, conical flask, measuring jar

CHEMICALS REQUIRED

Sodium Hydroxide, Manga nous Sulphate, Potassium iodide, Sodium Thiosulphate, Conc.H2SO4, Starch

REAGENT PREPARATION

1. Manga nous Sulphate

10gms of Manga nous Sulphate is dissolved in 25ml of distilled water.

2. Alkaline –Iodide Solution

2gms of Sodium Hydroxide and 3.75gms of Potassium iodide are dissolved in 25ml of distilled water.

3. Sodium Thio sulphate Solution (0.01N)

2.48gms of Sodium Thio Sulphate is dissolved in 1 liter of water.

4. Starch Solution

Take 1 gm of starch. Prepare paste with distilled water. Make 100 ml with water and boil by stirring and cool it.

5. Pipette Solution

2ml of Manga nous Sulphate solution and 2ml of alkaline Iodide Solution is added to 250ml of the sample taken in a reagent bottle. The bottle is stoppered and shaken thoroughly when the precipitate formed is settled, 2ml of Conc. H_2SO_4 is added and shaken thoroughly until the precipitate gets dissolved completely.

OBSERVATION

Source of sample:

Date of collection:

Time :

Temperature :

TABULATION

Sodium thiosulphate Vs. given sample

Sl. No.	Vol. of given sample (ml)	Burette Reading(ml)		Vol of $NO_2S_2O_3$ Consumed	Indicator	Endpoint
		Initial	Final			

1000 ml of N thiosulphate solution = 8gms of oxygen

Dissolved Oxygen in mg/L = V2*N*8*1000/V1

Where

V_1 = Volume of water sample in ml.

V_2 = Volume of Sodium thiosulphate in ml.

N = Normality of sodium thiosulphate

PROCEDURE

1. Take 50ml of clear pipette solution in a conical flask
2. Add to it one or two drops of starch indicator until the colour becomes blue.
3. Titrate against Standard Sodium Thiosulphate solution until the disappearance of colour.
4. Repeat the titration for concordant values.

SANITARY SIGNIFICANCE

In liquid wastes Dissolved Oxygen is the most important factor in determining whether aerobic or anaerobic organisms carryout biological changes. If sufficient D.O is available aerobic organisms oxidize the wastes to stable products. If D.O is deficient anaerobic bacteria take part in the conversion and reduce the waste often to obnoxious and nuisance conditions are usually resulted.

Application in Environmental Data:

1. It is one of the most important tests often used in most instances involving stream pollution control.
2. For the survival of aquatic life maintenance of D.O level is a must.
3. Determination of D.O serve as the basis of B.O.D test and thus they are the foundation of the most important determination used to evaluate pollution strength of sewage and industrial waste.

RESULT

Amount of Dissolved Oxygen present in a given sample is ----------------

TEST 13

Determination of Sulphate

OBJECTIVES

To determine the amount of sulphate present in the given sample by gravimetric method.

PRINCIPLE

The sulphate in water is precipitated as Barium Sulphate by the addition of Barium Chloride in hydrochloric acid medium. The precipitated is filtered, washed free of chloride, ignited and weighed as barium sulphate.

$Na_2SO_4 + BaCl_2 \longrightarrow BaSO_4 + 2NaCl$

$BaSO_4 + 2C \longrightarrow BaS + 2CO_2$

$BaS + 2O_2 \longrightarrow BaSO_4$

APPARATUS REQUIRED

Crucible, Oven, weighing balance, pipette, beaker, water bath, desiccator.

REAGENTS REQUIRED

Dilute HCL.

Solid Ammonium Chloride 10%Barium

Chloride solution.

Reagents Preparation:

  Barium Chloride Solution:

   Dissolve 10gm of Barium Chloride in 100 ml of distilled water.

PROCEDURE

1. Pipette out 50ml of sample into a clean 250ml beaker.
2. Add 10ml of dilute HCL & 1gm of solid ammonium chloride.

3. Heat to boiling &add 10ml of 10% Barium Chloride solution drop by drop with constant stirring. Continue boiling for another 2 to 3 minutes.
4. Allow the precipitate to settle and test for completion of precipitation by adding a small amount of Barium Chloride solution through the sides of the beaker.
5. If any turbidity is noticed add sufficient quantity of barium chloride to precipitate all the sulphate.
6. Transfer the contents to a sand bath &digest for half an hour to promote granulation of the precipitate.
7. Filter through Watt man no 42 filter paper and wash with boiling water till the filtrate runs free of chlorine.
8. Transfer the filter paper along with the precipitate to a weighed silica crucible and dry it an air oven.
9. Ignite over a low flame initially, taking care to ash the filter paper completely, then ignite strongly over a rose head flame to constant weight.
10. From the weight of Barium Sulphate obtained calculate the Sulphate content of the sample.

ENVIRONMENTAL SIGNIFICANCE

Sulphates in natural waters in concentrations ranging from a few to thousand mg/Excess Na_2SO_4 and $MgSO_4$ should not be present in drinking waters as they cause Cathartic action. Higher Concentration of Sodium Sulphate in water can cause malfunctioning of the alimentary canal. So the recommended upper limit is 250m/l in water intended for human consumption. In anaerobic decomposition of waste waters, Sulphates are reduced to hydrogen Sulphide causing obnoxious odours and promote corrosion of sewers. Sulphates are reduced to sulphide in sludge digesters and may upset the biological process, if the sulphide concentration exceeds 200mg/L.

Application of Sulphate Data in Environmental Engineering Practice:
1. The sulphate content of natural waters is an important consideration in determining their suitability for public and industrial water supplies.

2. The amount of sulphate in waste water is a factor of concern in determining the magnitude of problems that can arise from reduction of sulphates to hydrogen sulphide.
3. A knowledge of the sulphate content of the sludge or waste fed to digestion units provides a means of estimating the hydrogen sulphide content of the gas produced. From this information, the design engineer can determine whether scrubbing facilities will be needed to remove hydrogen sulphides and size of the units required.

OBSERVATION

Source of sample:
Date of collection:
Time   :
Temperature   :

TABULATION

Sl. No.	Weight of empty crucible(gm) A	Weight of crucible after heated in oven(gm) B

CALCULATION

Volume of sample taken = 50ml

Weight of empty silica crucible = A gm

Weight of crucible+precipitate = B gm

Weight of precipitate alone = (B-A) gm

233g of $BaSO_4$ contains 96g of SO_4

(B-A) g of $BaSO_4$ will contain = 96*(B-A)/233

In 10^6 ml = 96*(B-A)*10^6/ 233*50 ppm

RESULT

The amount of Sulphate present in the given sample _____ mg/L.

TEST   **14**

Determination of Fluorides

OBJECTIVE

To determine the fluorides present in the give sample.

APPARATUS REQUIRED

Burette with stand, Pipette, Conical flask, measuring flask.

CHEMICAL REQUIRED

Oxalate, concentrated hydrochloric acid, phenolphthalein indicator, sodium hydroxide.

REAGENTS PREPARATION

Oxalate Solution:

Dissolve 630mg of oxalate in distilled water and make up to 100ml.

Phenolphthalein indicator:

Add 1gm of phenolphthalein in 200 ml distilled water and dissolve it. Add 0.02N Sodium hydroxide solution drop wise until a faint pink colour appears.

Sodium hydroxide solution:

Dissolve 4gm of sodium hydroxide in distilled water and make up to 100ml.

PROCEDURE

Titration – I

NaoH Vs Oxalic acid

1. Pipette 20ml of oralic acid solution into the conical flash.
2. Add one or two drops of phenolphthalein indicator.
3. Titrate against sodium hydroxide solution until the appearance of pink colour.
4. Repeat the titration for concordant values.

Titration – II
NaoH Vs Sample:
1. Take 19ml of sample in the conical flask and add 1ml of concentrated hydrochloric acid.
2. Add one or two drops of phenolphthalein indicator.
3. Titrate against sodium hydroxide solution until the appearance of pink colour.
4. Repeat the titration for concordant values.

Titration – III
NaoH Vs Hcl:
1. Take 19ml of distilled water in the conical flask and add 1ml of concentrated hydrochloric acid.
2. Add one or two drops of phenolphthalein indicator.
3. Titrate against sodium hydroxide solution until the appearance of pink colour.
4. Repeat the titration for concordant values.

ENVIRONMENTAL SIGNIFICANCE

Fluorine in Human health
Presence of large amount of fluoride is associated with dental and skeletal fluorosis (>1.5 mg/l) and in adequate amounts with dental caries (<1mg/l).

Dental Fluorosis
In young children the diseases affects only on the teeth. This is known as dental fluorosis. The teeth lose their shiny appearance and chalk-white patches develop on them. This is known as mottled enamel and is an early sign of dental fluorosis. The white patches later become yellow and turn brown or black.

Skeletal Fluorosis
Inn aged people the disease affects the bones, tendons and ligaments. This is known as skeletal fluorosis. This is followed by pain and stiff of the back and later the joints of both limbs and limitation of neck movements.

Genu Valgum

It was observed that this syndrome was most prevalent among people whose staple diet was sorghum. Further studies have shown that diets based upon jower promote a higher retention of ingested fluoride than their based on rice.

Application of fluoride data in Environmental Engineering Practice:
1. Fluoride of water is an important in determining the suitability of water from domestic and industrial uses.
2. The size and design of Deflouridation units depends upon the level of fluorides present in the water.

OBSERVATION
Source of sample
Date of collection
Time
Temperature

Titration – I NaoH Vs. Oxalic acid

Sample details	Volume of Sample taken (ml)	Burette reading (ml)		Volume of NaoH used (ml)	End Point
		Initial reading	Final reading		
Oxalic acid					Appearance of pale pink colour

Titration – II   NaOH Vs Sample

Sample details	Volume of Sample taken (ml)	Burette reading (ml)		Volume of NaoH used (ml)	End Point
		Initial reading	Final reading		
Tap Water					Appearance of pale pink colour

Titration – III   NaOH Vs Distilled water

Sample details	Volume of Sample taken (ml)	Burette reading (ml)		Volume of NaoH used (ml)	End Point
		Initial reading	Final reading		
Distilled water					Appearance of pale pink colour

CALCULATION
From Titration – I,
   Normality of NaOH X = Vol. of oxalic acid * Normality of oxalic acid /Volume of NaOH

From Titration – II,
Normality of X = Vol. of NaOH * Normality of NaOH * 0.0091 /Volume of sample

Amount of fluorides = Vol. of NAOH in Titration (final-initial)*X*Y/ vol of sample

RESULT
   Amount of fluorides present in the given sample (Tap water) -------------mg/L.

TEST 15

Determination of Ammoniacal Nitrogen

OBJECTIVE
To determine the ammoniacal nitrogen present in the given sample.

APPARATUS REQUIRED
Burette with stand, Pipette, Conical flask, measuring flask.

CHEMICAL REQUIRED
Oxalate, concentrated hydrochloric acid, phenolphthalein indicator, sodium hydroxide, ammonia solution.

REAGENTS PREPARATION

Oxalate Solution:

Dissolve 630 mg of oxalate in distilled water and make up to 100 ml.

Phenolphthalein indicator:

Add 1 gm of phenolphthalein in 200 ml distilled water and dissolve it. Add 0.02 N Sodium hydroxide solution drop wise until a faint pink colour appears.

Sodium hydroxide solution:

Dissolve 4g of sodium hydroxide in distilled water and make up to 100 ml.

Standard Hydrochloric acid:

Dissolve 2 ml of HCl in distilled water and make up to 100 ml.

PROCEDURE

Titration – I NaOH Vs. Oxalic acid
1. Pipette 20 ml of oxalic acid solution into the conical flask
2. Add one or two drops of phenolphthalein indicator.
3. Titrate against sodium hydroxide solution until the appearance of pink colour.
4. Repeat the titration for concordant values.

Titration – II NaOH Vs. Ammonia
1. Take 17 ml of distilled water in the conical flask and add 1 ml of ammonia solution and 2 ml of hydrochloric acid.
2. Add one or two drops of phenolphthalein indicator.

CALCULATION

From Titration – I,

Normality of NaOH = Vol. of oxalic acid * Normality of oxalic acid (0.1) /Volume of NaOH

From Titration – II,

Normality of X = Vol. of NaOH * Normality of NaOH * 0.0091 /Volume of sample

Amount of Ammoniacal Nitrogen in mg/L = X *Vol of NaOH/ Volume of sample

3. Titrate against sodium hydroxide solution until the appearance of pink colour.
4. Repeat the titration for concordant values.

Titration – III   NaoH Vs Sample

1. Take 17 ml of distilled water in the conical flask and add 1 ml of ammonia solution and 2 ml of hydrochloric acid.
2. Add one or two drops of phenolphthalein indicator.
3. Titrate against sodium hydroxide solution until the appearance of pink colour.
4. Repeat the titration for concordant values.

ENVIRONMENTAL SIGNIFICANCE

1. Excess of ammonia in the form of nitrogen leads of Eutrophication in lakes.
2. Consumption of Nitrogen greater than 2 mg/l in drinking water may lead to mathemoglobonemia in children.

Application

1. Determination of ammoniacal nitrogen used for standardizing the drinking water supply.
2. The data is used in the treatment of waste water before it is subjected to water courses.
3. It is also used to determine the extent of eutrophication and possible methods of removal of Nitrogen.

RESULT

Amount of Ammonical Nitrogen present in the given sample is _____ mg/L

TEST 16

Determination of COD

OBJECTIVE
To determine the amount of Chemical Oxygen Demand present in the given sample.

PRINCIPLE
COD test is widely used for measuring the pollution strength of waste water. All organic compounds with a few exceptions can be oxidized to CO_2 and water by the action of strong oxidizing agents regardless of biological assimilability of the substances.

APPARATUS REQUIRED
1. COD reactor.
2. Burette with stand.
3. Pipette.
4. Measuring jar.
5. Tiles.
6. Beakers.
7. Conical flask.

CHEMICALS REQUIRED
1. Std. Potassium dichromate.
2. Conc. Sulphuric.
3. Ferroin indicator solution.
4. Std. Ferrous ammonium sulphate solution.
5. Mercuric Sulphate.

REAGENTS PREPARATION

Std. Potassium dichromate (0.25 N) Solution:

Dissolve 12.26 gm of potassium dichromate previously dried at 180^0C for 2 hr in distilled water and diluted to 1 liter.

Ferroin Indicator Solution:

Dissolve 1.485 gm of 1, 10 Phenophtholine sulphate monohydrate with 0.695gm of ferrous Sulphate ($FeSO^4.7H^2$) in water and dilute to 100 ml.

OBSERVATION:

Source of sample :

Date of collection :

Time :

Temperature :

TABULATION

Sl. No.	Sample details	Volume of Sample taken (ml)	Burette reading (ml)		Volume of NaOH used (ml)	End Point
			Initial reading	Final reading		

Concordant Value = _____ ml

CALCULATION

$$\text{COD mg/L} = \frac{(A - B) C \times 8 \times 1000}{\text{ml of sample used}}$$

Where:

A = ml of ferrous ammonium sulphate used for blank.

B = ml of ferrous ammonium sulphate used for sample.

C = Normality of ferrous ammonium sulphate solution.

Std. Potassium dichromate (0.25 N):

Dissolve 98 gm FAS in distilled water and add 20 ml of Conc.H_2SO_4. Cool and dilute to 1 liter. This solution must be standardized against the $K_2Cr_2O_7$ every day of its use.

PROCEDURE

1. Take 50 ml of sample in a flask and add boiling chips and 1 gm of $HgSO4$ and 5 ml of H_2SO_4 add slowly to dissolve $HgSO_4$ and cool the mixture.
2. Add 25ml of 0.25 N $K_2Cr_2O_7$ solution and gain mix. Attach the condense and start the cooling water. The remaining acid agent is added thoroughly through the open end of condenser and the efflux mixture was mixed. Apply the heat and reflux for 2 hrs.
3. Dilute the mixture to about 300 ml and titrate excess dichromate with std. FAS using Ferro in indicator.
4. The colour will change from yellow to green to blue and finally red and the ml of titrate was deduced.
5. Reflux on the same manner to flask consisting of distilled water, equal to the volume of the sample and the reagents titrate as he sample and ml of titrate was deduced.

ENVIRONMENTAL SIGNIFICANCE

1. BOD cannot be determined accurately when toxins are present and conditions are un favor for the growth of microbes.
2. BOD test consumes more time i.e. a minimum of 5 days where COD test is relatively faster than BOD taking only 3hr for completion.

Application of COD

1. COD test used extensively in the analysis of industrial wastes.
2. It is particularly valuable in survey system to determine and control losses in sewer system.
3. This test is widely used I BOD in the operation of treatment facilities because of the speed with which the result can be obtained.
4. It is useful to access the strength of waste which contains toxins and biological resultant and organic substance.
5. The ratio of BOD to COD is useful to Access the amenability of waste for biological treatment.
6. The ratio of BOD to COD is greater than or equal to 0.8 indicates that the waste water are highly amenable to biological treatment.

RESULT

The amount of COD present in the given sample is _____ mg/L.

TEST 17

Determination of Iron

OBJECTIVE
To estimate the amount of ferrous iron present in 100 ml of the given solution.

PRINCIPLE
The estimation is based on the redox reaction between ferrous sulphate and potassium permanganate

$$2KMnO_4 + 3H_2SO_4 \longrightarrow K_2SO_4 + 2MnSO_4 + 3H_2O + 5O$$

$$2FeSO_4 + 3H_2SO_4 + O \longrightarrow Fe_2(SO_4)_3 + H_2O$$

CHEMICALS REQUIRED

Std. 0.1023 Mohr's solution.

Potassium Permanganate Solution.

PROCEDURE

Titration I Standardization of Potassium Permanganate

1. A 50 ml burette is washed with water and rinsed with potassium permanganate solution.
2. It is then filled with potassium permanganate up to zero mark and the reading is noted.
3. Exactly 20 ml of the given Mohr's salt solution is pipette out into a clean conical flask. Then about one test tube full of dilute sulphuric acid is added.
4. This solution is titrated against potassium permanganate taken in the burette. The end point is the appearance of pale permanent pink color.
5. The titration is repeated for concordant values.

Titration II Standardization of Potassium Permanganate

1. A 50 ml burette is washed with water and rinsed with potassium permanganate solution.
2. It is then filled with potassium permanganate up to zero mark and the reading is noted.

OBSERVATION

Source of sample :
Date of collection :
Time :
Temperature :

TABULATION

Std. FAS Vs link $KMnO_4$

Sl. No.	Sample details	Volume of Sample taken (ml)	Burette reading (ml)		Volume of NaOH used (ml)	End Point
			Initial reading	Final reading		

Concordant Value =   ml

CALCULATION

Volume of FAS V1   =   ml
Strength of FAS N1   =
Volume of $KMnO_4$ V2   =   ml
Strength of $KMnO_4$ N2 =   ?
V1*N1   =   V2*N2
N2   =   V1*N1 / V2

Link $KMnO_4$ Vs Ferrous Ion Solution

Sl. No.	Sample details	Volume of Sample taken (ml)	Burette reading (ml)		Volume of NaOH used (ml)	End Point
			Initial reading	Final reading		

Concordant Value =   ml

CALCULATION

Volume of $KMnO_4$ V1   =   ml
Strength of $KMnO_4$ N1   =
Volume of Ferrous Iron solution V2   =   ml
Strength of Ferrous Iron solution N2 =   ?

V1*N1   =   V2*N2
N2   =   V1*N1 / V2

Amount of ferrous iron solution per liter   =   *Normality * Equivalent weight/*

10

Equivalent Weight of ferrous iron   =   55.84

3. Exactly 20 ml of the made up ferrous iron solution is pipette out into a clean conical flask. Then about one test tube full of dilute sulphuric acid is added.
4. This solution is titrated against potassium permanganate taken in the burette. The end point is the appearance of pale permanent pink color.
5. The titration is repeated for concordant values.

RESULT

Amount of ferrous Iron in 100 ml of the given sample = _____ gm

TEST 18

Determination of Biochemical Oxygen Demand

GENERAL

Biochemical Oxygen Demand (BOD) is defined as the amount of oxygen required by bacteria for stabilizing decomposable organic matter in water under aerobic conditions. BOD is expressed in milligram per liter.

OBJECTIVE

To determine the amount of BOD in the given Sewage Sample.

PRINCIPLE

An amount of the water sample is maintained in an incubator at 20^0C for five days in a closed bottle without allowing air to enter during which time the water sample is assumed to be sterilized, ie the bacterial decomposition gets completed. Measuring the dissolved oxygen in the water sample before and after incubation would incubate the amount of oxygen used for stabilizing the water.

A water sample with a low BOD can be straight away used for the BOD determination. But a water sample with a high BOD must be diluted and pretreated before the determination of BOD.

APPARATUS REQUIRED

Burette with stand, Pipette, Conical flask, measuring jar.

REAGENTS PREPARATION

Phosphate Buffer Solution:
Dissolve 8.5gm KH_2PO_4, 21.75gm K_2HPO_4, 33.4gm $Na_2HPO_47H_2$) & 1.7gm NH_4Cl in about 500 ml distilled water and dilute to 1.0 liter. The pH of this buffer solution should be 7.2 without further adjustment.

Ferric chloride Solution:
Dissolve 0.25 gm Ferric chloride ($FeCl_3$ $6H_2O$) in distilled water and dilute to 1.0 liter.

Calcium Chloride Solution:
Dissolve 27.5 gm of calcium chloride per liter of distilled water.

Magnesium Sulphate Solution:
Dissolve 2.26gm of magnesium sulphate ($MgSO_47H_2O$) in distilled water and dilute to 1.0 liter.

OBSERVATION

    Source of sample   : Sewage
    Date of Collection   :
    Time   :
    Temperature   :

TABULATION

Sl. No.	Volume of the given sample (ml)	Burette Reading (ml)		Indicator	End point
		Initial	Final		

CALCULATION

1000 ml of 1N thiosulphate solution = 8 gm of oxygen
Dissolved oxygen in mg/L= $V_2 * N * 8 * 1000 * V_1$

Where,
    V_1 = Volume of sample

    V_2 = Volume of Sodium thiosulphate in ml

    N = Normality of Sodium thiosulphate

Before Incubation

Dissolved oxygen in mg/L= $V_2 * N * 8 * 1000 * V_1$

After Incubation

Dissolved oxygen in mg/L= $V_2 * N * 8 * 1000 * V_1$

WATER FOR DILUTION

Take 2 liter of distilled water in a 3 liter bottle. Shake this partially filled bottle for about ten minutes. So that the distilled water gets saturated with atmospheric oxygen. Then add to this 2ml of phosphate buffer solution, 2ml of magnesium sulphate, 2ml of calcium chloride solution and 2ml of ferric chloride solution.

PROCEDURE

BOD Determination of without Dilution

If the water sample is fairly clean would have a BOD value of less than 5 and it can be used as such Fill 4 BOD bottles up to the brim. After filtration fifteen minutes tightly stopper these. Make sure that these are no air bubbles. Use the water in two bottles and determine the dissolved oxygen immediately place the other two bottles in a incubator or a constant temperature water bath at 20^0. During incubation the bottles should be protected from entry of air into them. This can be done by keeping the bottles inverted so that the bottles are inside water in a tray. After five days the dissolved oxygen in these two bottles are determined.

BOD (mg/L) = D.O before incubation – D.O. after incubation.

BOD determination after dilution:

This method is used when the BOD value exceed 5 pipette into each of two BOD bottles 100 ml water sample. The bottles are filled with the dilution bottle prepared. Wait for 15 minutes and stopper the bottle tightly. The DO of the diluted water using one of the bottles is determined immediately. The other bottles are incubated for 5 days at 20^0C and then the DO of the incubated water is determined.

TEST 19

Introduction to Bacteriological Analysis

OBJECTIVE

To study the total bacterial count and MPN of coliforms present in a given Sample.

PROCEDURE

1) **STANDARD PLATE COUNT TEST:**

   1. Select the dilution ratios depend upon expected total bacterial count (1:0, 1:100, 1:1000).
   2. A separate sterile pipette should be used for each dilution.
   3. Transfer 1 ml, from undiluted and 0.1 ml and 0.01 ml from diluted samples in sterile Petri dish. After delivering the sample, touch the tip of pipette to a dry spot in the plates.
   4. Pour 10 ml of sterile nutrient agar medium of the temp $44 - 46^0C$ (high temperature may kill bacteria) to these Petri dishes by gently opening dish plates slightly.
   5. Mix the medium thoroughly with sample in Petri plates. When the media is solidified, invert the plates and keep for incubation at $37^0C \pm 5^0C$ for $44 \pm 4h$.
   6. The visible colonies are then counted with the aid of $6 - 8$ X magnify glass or colony counter.

SPC/ml= colonies counted / Dilution factor

Environmental Significance

The total bacterial count is the number of visible colonies under a magnification of $6 - 8$ X which have developed under defined conditions. It provides a measure of the degree of microbiological contamination of the water and especially of sudden bacterial invasions.

Its values are useful in warning about excessive, microbial growth in any water and also in judging the efficiency of water and waste water treatment in removing organisms.

2) **MOST PROBABLE NUMBER TEST**:

Shake the water sample thoroughly before making dilutions or before inoculation.

a) Presumptive test:

Type of water	Dilution and tubes
Potable waters	5 tubes of 10 ml double strength medium with 1 ml sample
	5 tubes of 10 ml double strength medium with 10 ml sample
	5 tubes of 10 ml single strength medium with 1 ml sample
Raw waters	5 tubes of 10 ml single strength medium with 0.1 ml sample
	5 tubes of 10 ml single strength medium with 0.01 ml sample
	5 tubes of 10 ml single strength medium with 0.01 ml sample
Polluted water	5 tubes of 10 ml single strength medium with 0.001 ml sample

1. Select dilutions according to the expected coliforms
2. Inoculate a series of MPN fermentation tube with appropriate measured quantities of the water sample to be tested.
3. Add appropriate quantity of Lauryl try lose broth in each tube.
4. Put one Durham's vial inverted in each test tube. The top of the tubes is plugged with cotton plug.
5. Place all these tubes in an incubator at $35\text{-}37^0C$ within 30 minutes.
6. After 24 hrs, examine the tubes carefully. Those showing gas in the Durham's vial are recorded as positive (+). If no gas has been formed, reincubate for another 24 hrs and at the end of 48 hrs, examine again.
7. Record the presence or absence of gas at each examination of the tubes regardless of the amount.
8. Formation of the gas within 48 ± 3 hrs of incubation is considered as negative test.

b) Confirmed test:

1. Prepare fermentation tubes with 10 ml brilliant green lactose bile broth and Durham's vials. The number of tubes to be prepared is equal to be prepared is equal to all positive tests in the presumptive test.
2. Shake gently the fermentation tubes with positive results and transfer one loop or two loopful of medium to brilliant green lactose bile broth by using a sterile metal loop.
3. Incubate the tubes at $35 – 37^0C$ for 48 ± 3 hrs.
4. The formation of gas in any amount in the Durham's vials of BGLB tubes at any time within 48 ± 3 hrs constitutes a positive confirmed test.

c) Completed Test:

The completed test through not always necessary is sometimes done to demonstrative with certainty that organisms showing positive results for the confirmed test are really members of the coliform group.

1. Prepare Endo or Eosin Methylene Blue (EMB) agar Petri plates. The number of Petri plates to be prepared is the same as that of tubes showing gas production in BGLB medium.
2. Streak inoculum from the BGLB tubes on these plates in such a way that these plates in such a way that the colonies after separation have a distance of 0.5 cm.
3. Incubate these plates at $37^0C \pm 0.2$ for 24 ± 2 hrs.
4. Examine the plates or bacterial growth as colonies. Well isolated colonies with a dark center (nucleated) are typical coliform colonies. They may have a metallic surface sheen. The colonies that are pink or opaque and are not nucleated are typical colonies and may belong to the coliform group. Clear, watery colonies are not of coliform and are reported as negative in the completed test.
5. Now inoculate an isolated coliform colony (avoid picking up a mixture) from each plate into the tubes a lauryl trytose broth and record the gas production within 48 hrs at 37^0C.

Computing and Recording of MPN

Record the number of positive finding of coliform group organisms (presumptive, confirmed and completed) resulting from multiple portion decimal dilution planting as the combination of positives and compute in terms of the MPN.

Environmental Significance

The purpose of this test is to estimate the number of coliforms in water sample as an index of the magnitude of biological contamination. The intestinal tract of human beings contains countless rod shaped bacteria known as Coliform organisms. Each person discharges 100 to 400 billion coliform organisms per day in addition to other kinds of bacteria. Coliforms are harmless to ma and are, in fact, useful in destroying organic matter in biological waste treatment processes.

Pathogenic organisms present in wastes and polluted water are few, very sensitive to slight change of environmental factors and difficult to isolate, whereas the coliform are more numerous, more easily tested and thus used as an indicator organism.

The presence of coliform organisms are taken as an indication that pathogenic organism may also present and the absence of coliform organisms.

Application of MPN Data in Environmental Engineering Practice

1. MPN determination is useful in the selection of water supplies for human needs.
2. It is useful to assess the degree of contamination or receiving waters due to domestic sewage.
3. It is useful to find efficiency of disinfectant and treatment unit's efficiency in
4. Pathogenic bacterial removal.

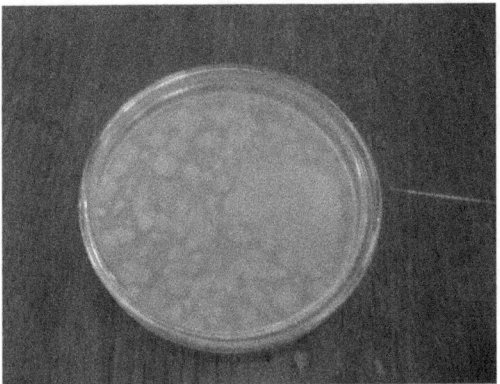

Fig. Presence of Bacteria

RESULT:

Thus the plate count and MPN test methods for bacteriological analysis are studied.

Definitions:

A

Abatement is the reduction or elimination of the degree or intensity of emissions i.e. pollution.

Abiotic Resources are the resources which are considered abiotic and therefore not renewable. Zinc ore and crude oil are examples of abiotic resources.

Acceptable Daily Intake is the highest daily amount of a substance that may be consumed over a lifetime without adverse effects.

Acid Deposition is a comprehensive term for the various ways acidic compounds precipitate from the atmosphere and deposit onto surfaces. It can include:
- wet deposition by means of acid rain, fog, and snow
- dry deposition of acidic particles (aerosols).

Acid Rain is rain mixed mainly with nitric and sulphuric acid that arise from emissions released during the burning of fossil fuels.

Acute Exposure is one or a series of short-term exposures generally lasting less than 24 hours.

Adaptability refers to the degree to which adjustments are possible in practices, processes, or structures of systems to projected or actual changes of climate. Adaptation can be spontaneous or planned, and be carried out in response to or in anticipation of changes in conditions.

Aerobic composting is a method of composting organic waste using bacteria that need oxygen. This requires that the waste be exposed to air either by turning or by forcing air through pipes that pass through the material.

Aerosols are particles of solid or liquid matter that can remain suspended in air from a few minutes to many months depending on the particle size and weight.

Air is a mixture of gases containing about 78 per cent nitrogen; 21 percent oxygen; less than 1 percent of carbon dioxide argon, and other gases; and varying amounts of water vapor.

Air Monitoring is the sampling for and measuring of pollutants present in the atmosphere.

Air Pollution is the degradation of air quality resulting from unwanted chemicals or other materials occurring in the air.

Air Pollutants are amounts of foreign and/or natural substances occurring in the atmosphere that may result in adverse effects to humans, animals, vegetation, and/or materials.

Air Quality Standard (AQS) is the prescribed level of a pollutant in the outside air that should not be exceeded during a specific time period to protect public health.

Alternative Fuel are fuels such as methanol, ethanol, natural gas, and liquid petroleum gas that are cleaner and help to meet mobile and stationary emission standards. These fuels may be used in place of less clean fuels for powering motor vehicles.

Ambient Air is the air occurring at a particular time and place outside of structures. Often used interchangeably with outdoor air.

Ambient Air Quality Standards (AAQS) are health and welfare-based standards for outdoor air which identify the maximum acceptable average concentrations of air pollutants during a specified period of time.

Ammonia is a pungent colorless gaseous compound of nitrogen and hydrogen that is very soluble in water and can easily be condensed into a liquid by cold and pressure. Ammonia reacts with NO_x to form ammonium nitrate.

Asbestos is a mineral fiber that can pollute air or water and cause cancer or asbestosis when inhaled. The U.S. EPA has banned or severely restricted its use in manufacturing and construction and the ARB has imposed limits on the amount of asbestos in serpentine rock that is used for surfacing applications.

Atmosphere is the gaseous mass or envelope of air surrounding the Earth. From ground-level up, the atmosphere is further subdivided into the troposphere, stratosphere, mesosphere, and the thermosphere.

Aquaculture or pisceculture is the breeding or rearing of freshwater or marine fish in captivity, fish farming.

B
Binding targets refers to environmental standards that are to be met in the future.
Biodegradable material is any organic material that can be broken down by microorganisms into simpler, more stable compounds. Most organic waste such as foods, paper, etc. are biodegradable.
Biogenic Source are biological sources such as plants and animals that emit air pollutants such as volatile organic compounds Examples of biogenic sources include animal management operations, and oak and pine tree forests.
Biomass is the living materials (wood, vegetation, etc.) grown or produced expressly for use as fuel.
Biomass burning is the burning of organic matter for energy production, forest clearing and agricultural purposes. Carbon dioxide is a bi-product of biomass burning
Biomass fuels are wood and forest residues, animal manure and waste, grains, crops and aquatic plants are some common biomass fuels.
Biome is a climatic region characterized by its dominant vegetation.
Bio reserves are the areas with rich ecosystems and species diversity are reserved for conservation.
Biota is the flora and fauna of an area.
Biotic are the resources which are considered biotic and therefore renewable. The rainforests and tigers are examples of biotic resources.
BOD is the biochemical oxygen demand.
Brackish water contains 500 to 3000ppm of sodium chloride.

C
Calorie Metric thermal unit is a measure of heat energy; the amount needed to raise the temperature of one kilogram of water by one degree Centigrade. This is the large Calorie (used relating to food energy content) definition. The "small" calorie of fuel research is the amount of energy needed to raise the temperature of one gram of water by one degree Centigrade.
Carbon cycle is the process of removal and uptake of carbon on a global scale. This involves components in food chains, in the atmosphere as carbon dioxide, in the hydrosphere and in the geosphere. The major movement of carbon results from photosynthesis and from respiration. See also sink and source.
Carbon Dioxide (CO_2) is a colourless, odourless gas that occurs naturally in the Earth's atmosphere. Significant quantities are also emitted into the air by fossil fuel combustion and deforestation. It is a greenhouse gas of major concern in the study of global warming. It is estimated that the amount in the air is increasing by 0.27% annually.
Carbon Monoxide (CO) is a colourless, odourless gas resulting from the incomplete combustion of hydrocarbon fuels. CO interferes with the blood's ability to carry oxygen to the body's tissues and results in numerous adverse health effects. Over 80% of the CO emitted in urban areas is contributed by motor vehicles. CO is a criteria air pollutant.
Carbon sequestration generally refers to capturing carbon -- in a carbon sink, such as the oceans, or a terrestrial sink such as forests or soils -- so as to keep the carbon out of the atmosphere.
Carbon sink is a pool (reservoir) that absorbs or takes up released carbon from another part of the carbon cycle. For example, if the net exchange between the biosphere and the atmosphere is toward the atmosphere, the biosphere is the source, and the atmosphere is the sink.
Carnivore is the flesh eating species.
Carrying capacity is the maximum number of organisms that can use a given area of habitat without degrading the habitat and without causing social stresses that result in the population being reduced.

Catalyst is a substance that can increase or decrease the rate of a chemical reaction between the other chemical species without being consumed in the process.

Catalytic converter is a motor vehicle pollution control device designed to reduce emissions such as oxides of nitrogen hydrocarbons carbon monoxide. Catalytic converters have been required equipment on all new motor vehicles sold in India.

Chlorofluorocarbons (CFCs) are synthetically produced compounds containing varying amounts of chlorine, fluorine and carbon. Used in industrial processes, refrigeration and as a propellant for gases and sprays. In the atmosphere they are responsible for the depletion of ozone and can destroy as many as 10,000 molecules of ozone in their long lifetime. Their use is now currently restricted under the Montreal Protocol.

Chronic health effect is a health effect that occurs over a relatively long period of time (e.g., months or years).

Climate is the prevalent long term weather conditions in a particular area. Climatic elements include precipitation, temperature, humidity, sunshine and wind velocity and phenomena such as fog, frost, and hail storms.

Climate change can be caused by an increase in the atmospheric concentration of greenhouse gases which inhibit the transmission of some of the sun's energy from the earth's surface to outer space. These gases include carbon dioxide, water vapor, methane, chlorofluorocarbons (CFCs), and other chemicals. The increased concentrations of greenhouse gases result in part from human activity -- deforestation; the burning of fossil fuels such as gasoline, oil, coal and natural gas; and the release of CFCs from refrigerators, air conditioners, etc.

COD is the chemical oxygen demand.

Combustion is the act or instance of burning some type of fuel such as gasoline to produce energy. Combustion is typically the process that powers automobile engines and power plant generators.

Community is a group of organisms living in a common environment and interdependent.

Compost is the material resulting from composting, which is the natural process of decomposition of organic waste that yields manure or compost, which is very rich in nutrients. Compost, also called humus, is a soil conditioner and a very good fertilizer.

Concentration is the measure of the atmospheric content of a gas, defined in terms of the proportion of the total volume that it accounts for. Greenhouse gases are trace gases in the atmosphere and are usually measured in parts per million by volume (ppmv), parts per billion by volume (ppbv) or parts per trillion (million million) by volume (pptv).

Conservation is the planning and management of resources to secure their long term use and continuity and better their quality, value and diversity. It is the use of less energy, either by using more efficient technologies or by changing wasteful habits.

D

Deforestation is the practice or process that results in the long-term change in land-use to non-forest uses. This is often cited as one of the major causes of the enhanced greenhouse effect for two reasons:
- The burning or decomposition of the wood releases carbon dioxide
- Trees that once removed carbon dioxide from the atmosphere in the process of photosynthesis are lost.

Depletion is the result of the extraction of abiotic resources (non-renewable) from the environment or the extraction of biotic resources (renewable) faster than they can be renewed.

Desertification is the progressive destruction or degradation of existing vegetative cover to form desert. This can occur due to overgrazing, deforestation, drought and the burning of extensive areas. Once formed, desert can only support a sparse range of vegetation. Climatic effects associated with this phenomenon include increased albedo, reduced atmospheric humidity and greater atmospheric dust loading, which can cause wind erosion and/or atmospheric pollution.

Diversity is the number of species in an area i.e. a community has a high degree of diversity if it contains many species of equal abundance.

E

Ecology is the study of the interrelationships between and among organisms and environment.

Efficiency is the ration of desired work-type output to the necessary energy input, in any given energy transformation divide. An efficient LIGHT bulb for example uses most of the input electrical energy to produce light, not heat. An efficient HEAT bulb uses most of its input to produce heat, not light.

El Niño is a climatic phenomenon occurring every 5 to 7 years during Christmas (El Niño means Christ child) in the surface oceans of the SE Pacific. The phenomenon involves seasonal changes in the direction of Pacific winds and abnormally warm surface ocean temperatures. The changes normally only effect the Pacific region, but major events can disrupt weather patterns over much of the globe. The relationship between these events and global weather patterns are poorly understood and are currently the subject of much research.

Emission is the release of a substance (usually a gas when referring to the subject of climate change) into the atmosphere.

Emission factor is the relationship between the amount of pollution produced and the amount of raw material processed or burned. For mobile sources, the relationship between the amount of pollution produced and the number of vehicle miles travelled. By using the emission factor of a pollutant and specific data regarding quantities of materials used by a given source, it is possible to compute emissions for the source. This approach is used in preparing an emissions inventory.

Endangered species are the plant and animal species in danger of extinction.

Endemic species are the species which are native, restricted or peculiar to an area.

Energy-efficient is electrical lighting devices which produce the same amount of light (lumens) using less electrical energy than incandescent electric light bulbs. Such devices are usually of the fluorescent type, which produce little heat, and may have reflectors to concentrate or direct the light output.

Energy efficiency is the amount of fuel needed to sustain a particular level of production or consumption, in an industrial or domestic enterprise. Energy efficiency measures are designed to reduce the amount of fuel consumed, either through greater insulation, less waste, or improved mechanical efficiencies, without losing any of the value of the product or process. Improving energy efficiency is a technological means to reduce emissions of greenhouse gases without increasing production costs.

Environment is the surroundings in which an organization operates, including air, water, land, natural resources, flora, fauna, humans, and their interrelations. This definition extends the view from a company focus to the global system.

Environmental effect is any direct or indirect impingement of activities, products and services of an organization upon the environment, whether adverse or beneficial. An environmental effect is the consequence of an environmental intervention in an environmental system.

Environmental impact is any change to the environment, whether adverse or beneficial, wholly or partially resulting from an organization's activities, products or services. An environmental impact addresses an environmental problem.

Estuary is a region where fresh water from a river mixes with salt water from the sea.

Ethanol is Ethyl-alcohol, a volatile alcohol containing two carbon groups. For fuel use, ethanol is produced by fermentation of corn or other plant products.

Evaporative emissions are the emissions from evaporating gasoline, which can occur during vehicle refueling, vehicle operation, and even when the vehicle is parked. Evaporative emissions can account for two-thirds of the hydrocarbon emissions from gasoline-fuelled vehicles on hot summer days.

Exposure is the concentration of the pollutant in the air multiplied by the population exposed to that concentration over a specified time period.

F

Fauna is the total animal life in an area.

Flora is the total plant life in an area.

Fluorescent light is a device which uses the glow discharge of an electrified gas for the illuminating element rather than an electrically heated glowing conductive filament.

Fly ash is air-borne solid particles that result from the burning of coal and other solid fuel.

Food chain is a sequence of organisms through which energy is transferred from its ultimate source in a green plant; the predator-prey pathway in which organism eats the next link below and is eaten by the link above.

Food web is a group of interconnecting food chains.

Fossil fuel is any hydrocarbon deposit that can be burned for heat or power such as coal, oil or natural gas. Fossil fuels are formed from the decomposition of ancient animal and plant remains. A major concern is that they emit carbon dioxide into the atmosphere when burnt, a major contributor to the enhanced greenhouse effect.

Or

Fossil fuels are the fuels formed eons ago from decayed plants and animals. Oil, coal and natural gas are such fuels.

Or

Fossil fuels such as coal, oil, and natural gas are so-called because they are the remains of ancient plant and animal life.

Fuel is a material which is consumed, giving up its molecularly stored energy which is then used for other purposes. e.g. to do work (run a machine).

Fuel efficiency is the amount of work obtained for the amount of fuel consumed. In cars, an efficient fuels allows more miles per gallon of gas than an inefficient fuel.

Fuel cell is an electrochemical cell, which captures the electrical energy of a chemical reaction between fuels such as liquid hydrogen and liquid oxygen and converts it directly and continuously into the energy of a direct electrical current.

Fumes are solid particles under 1 micron in diameter formed as vapor's condense, or as chemical reactions take place.

Furnace is combustion chamber; an enclosed structure in which fuel is burned to heat air or material.

G

Garbage is the waste that is generated whether in the household, commercial areas, industries, etc.

Gene is a section of a chromosome containing enough DNA to control the formation of a protein; a gene controls the transmission of a hereditary character.

Geothermal is pertaining to heat energy extracted from reservoirs in the earth's interior, as is the use of geysers, molten rock and steam spouts.

Geothermal energy is the heat generated by natural processes within the earth. Chief energy resources are hot dry rock, magma (molten rock), hydrothermal (water/steam from geysers and fissures) and geopressure (water satured with methane under tremendous pressure at great depths).

Global warming is an increase in the temperature of the Earth's troposphere. Global warming has occurred in the past as a result of natural influences, but the term is most often used to refer to the warming predicted by computer models to occur as a result of increased emissions of greenhouse gases.

Greenhouse effect is the progressive, gradual warming of the earth's atmospheric temperature, caused by the insulating effect of carbon dioxide and other greenhouse gases that have proportionately increased in the atmosphere. The greenhouse effect disturbs the way the Earth's

climate maintains the balance between incoming and outgoing energy by allowing short-wave radiation from the sun to penetrate through to warm the earth, but preventing the resulting long-wave radiation from escaping back into the atmosphere.
The heat energy is then trapped by the atmosphere, creating a situation similar to that which occurs in a car with its windows rolled up.
Greenhouse gases (GHGs) include the common gases of carbon dioxide and water vapor, but also rarer gases such as methane and chlorofluorocarbons (CFCs) whose properties relate to the transmission or reflection of different types of radiation. The increase in such gases in the atmosphere, which contributes to global warming, is a result of the burning of fossil fuels, the emission of pollutants into the atmosphere, and deforestation.

H
Habitat is the natural area in which a species or organism is found.
Hazardous waste is waste that is reactive, toxic, corrosive, or otherwise dangerous to living things and to the environment. Many industrial by products are hazardous.
Haze (Hazy) is a phenomenon that results in reduced visibility due to the scattering of light caused by aerosols. Haze is caused in large part by man-made air pollutants.
Herbivore is an animal that eats plants or parts of plants.
Hydro is that which is produced by or derived from water or the movement of water, as in hydroelectricity.
Hydrocarbons are compounds containing various combinations of hydrogen and carbon atoms. They may be emitted into the air by natural sources (e.g., trees) and as a result of fossil and vegetative fuel combustion, fuel volatilization, and solvent use. Hydrocarbons are a major contributor to smog.

I
Incineration is the process of burning solid waste and other material, under controlled conditions, to ash.
Indoor air pollution occur within buildings or other enclosed spaces, as opposed to those occurring in outdoor, or ambient air. Some examples of indoor air pollutants are nitrogen oxides, smoke, asbestos, formaldehyde, and carbon monoxide.
Inorganic waste is waste consisting of materials other than plant or animal matter, such as sand, glass, or any other synthetics.
Insolation is the solar radiant energy received by the earth.

J
Joint implementation is a concept where industrialized countries meet their obligations for reducing their greenhouse gas emissions by receiving credits for investing in emissions reductions in developing countries.

L
Leachate is the liquid that has seeped through a landfill or a compost pile. If uncontrolled it can contaminate both ground water and surface water.
Lead is a grey-white metal that is soft, malleable, ductile, and resistant to corrosion. Sources of lead resulting in concentrations in the air include industrial sources and crustal weathering of soils followed by fugitive dust emissions. Health effects from exposure to lead include brain and kidney damage and learning disabilities. Lead is the only substance that is currently listed as both a criteria air pollutant and a toxic air contaminant.

M

Methane (CH₄) is a greenhouse gas, consisting of four molecules of hydrogen and one of carbon. It is produced by anaerobically decomposing solid waste at landfills, paddy fields, etc.

Migration is the regular movements of animals, often between breeding places and winter feeding grounds.

Mudflats are area of mud that do not support any vegetation and are often covered by water.

N

Natural resources include renewable (forest, water, soil, wildlife, etc.) and non-renewable (oil, coal, iron ore etc.) resources that are natural assets.

Natural sources are the non-manmade emission sources, including biological and geological sources, wildfires, and windblown dust.

Nitrogen oxides (Oxides of Nitrogen, No_x) is a general term pertaining to compounds of nitric oxide (NO), nitrogen dioxide and other oxides of nitrogen. Nitrogen oxides are typically created during combustion, combustion processes, and are major contributors to smog formation and acid deposition. NO_2 is a criteria air pollutant and may result in numerous adverse health effects. They are produced in the emissions of vehicle exhausts and from power stations.

Nitrous oxide (N_2O) is a greenhouse gas, consisting of two molecules of nitrogen and one of oxygen.

O

Organic Compounds are a large group of chemical compounds containing mainly carbon, hydrogen, nitrogen, and oxygen. All living organisms are made up of organic compounds.

Organic waste is the material that is more directly derived from plant and animal sources, which can generally be decomposed by microorganisms.

Organisms are living thing, animal or plant that is capable of carrying out life processes.

OTEC - Ocean Thermal Energy Conversion Technology, which uses the temperature differential between warm surface water and cold deep water to run heat engines to produce electrical power.

Oxidant is a substance that brings about oxidation in other substances. Oxidizing agents (oxidants) contain atoms that have suffered electron loss. In oxidizing other substances, these atoms gain electrons. Ozone, which is a primary component of smog, is an example of an oxidant.

Oxidation is the chemical reaction of a substance with oxygen or a reaction in which the atoms in an element lose electrons and its valence is correspondingly increased.

Ozone (O3) it consists of three atoms of oxygen bonded together in contrast to normal atmospheric oxygen which consists of two atoms of oxygen. Ozone is formed in the atmosphere and is extremely reactive and thus has a short lifetime. In the stratosphere ozone is both an effective greenhouse gas (absorber of infra-red radiation) and a filter for solar ultra-violet radiation. Ozone in the troposphere can be dangerous since it is toxic to human beings and living matter. Elevated levels of ozone in the troposphere exist in some areas, especially large cities as a result of photochemical reactions of hydrocarbons and nitrogen oxides, released from vehicle emissions and power stations.

Ozone depletion is the reduction in the stratospheric ozone layer. Stratospheric ozone shields the Earth from ultraviolet radiation. The breakdown of certain chlorine and/or bromine-containing compounds that catalytically destroy ozone molecules in the stratosphere can cause a reduction in the ozone layer.

Ozone layer is the ozone in the stratosphere is very diffuse, occupying a region many kilometers in thickness, but is conventionally described as a layer to aid understanding.

P

Parasite is an organism that lives upon and at the expense of another organism.

Particulate matter (PM) is any material, except pure water, that exists in the solid or liquid state in the atmosphere. The size of particulate matter can vary from coarse, wind-blown dust particles to fine particle combustion products.
Percolation is the movement of water downwards and radially through the subsurface soil layers, usually continuing downward to the ground water.
Poaching is illegal hunting.
Pollution is the residual discharges of emissions to the air or water following application of emission control devices.
Population is a group of closely related and interbreeding organisms.
Precipitation is any or all form of liquid or solid water particles that fall from the atmosphere and reach the earth's surface. It includes drizzle, rain, snow and hail.
Predator is a animal that feeds on other animals.
Prey is an animal that is eaten by another animal.
Propellant is a gas with a high vapor pressure used to force formulations out of aerosol
spray cans. Among the gases used are butanes, propane's and nitrogen ozone hydrocarbons nitrogen oxides, and other chemically reactive compounds which, under certain conditions of weather and sunlight, may result in a murky brown haze that causes adverse health effects. The primary source of smog in California is motor vehicle.
Protected area is any area of land that has legal measures limiting human use of the plants and animals within that area; it includes national parks, game reserves, biosphere reserves, etc.

R
Range is the portion of the earth in which a given species is found.
Recharge is the process by which water is added to a reservoir or zone of saturation, often by runoff or percolation from the soil surface.
Recycling is the process of transforming materials (mainly waste) into raw materials for manufacturing new products.
Renewable energy is the energy resource that does not use exhaustible fuels. It is the energy from sources that cannot be used up: sunshine, water flow, wind and vegetation and geothermal energy, as well as some combustible materials, such as landfill gas, biomass, and municipal solid waste.
Resources are the materials found in the environment that can be extracted from the environment in an economic process. There are abiotic resources (non-renewable) and biotic resources (renewable).
Reservoir is any natural or artificial holding area used to store, regulate, or control a substance.
Runoff is that part of precipitation, snow or ice melt or irrigation water that flows from the land to the streams or other water surfaces.
Reuse is when we can use a product more than once in its original form.

S
Salinity is the degree of salt in the water or soil.
Smoke is a form of air pollution consisting primarily of particulate matter (i.e., particles released by combustion. Other components of smoke include gaseous air pollutants such as hydrocarbons oxides of nitrogen, and carbon monoxide. Sources of smoke may include fossil fuel combustion, agricultural burning, and other combustion processes.
Sulfur dioxide (SO_2) is a strong smelling, colorless gas that is formed by the combustion of fossil fuels. Power plants, which may use coal or oil high in sulfur content, can be major sources of SO_2. SO_2 and other sulfur oxides contribute to the problem of acid deposition. SO_2 is a criteria air pollutant.
Surface water is all water naturally open to the atmosphere.
Sustainable development implies economic growth together with the protection of environmental quality, each reinforcing the other. The essence of this form of development is a stable relationship

between human activities and the natural world, which does not diminish the prospects for future generations to enjoy a quality of life at least as good as our own.

Swamp is an area that is saturated with water for much of the time but in which soil surface is not deeply submerged.

Symbiosis is the living together in more or less close association of two dissimilar organisms, in which one or both derive benefit from the relationship.

T

TDS is the total dissolved solids.

Terrestrial is that which is of, or related to the land.

Tidal marsh is a low, flat, marshland traversed by inter laced channels and subject to tidal inundation. The only vegetation present is halo-tolerant bushes and grasses.

Turbidity is the cloudiness of a liquid caused by suspended matter.

V

Vapor is the gaseous phase of liquids or solids at atmospheric temperature and pressure.

Vertebrate is any of a major group of animals (fish, amphibians, reptiles, birds and mammals) with a segmented spinal column (backbone).

Volatile organic compounds(VOCs) are the carbon-containing compounds that evaporate into the air (with a few exceptions). VOCs contribute to the formation of smog and/or may themselves be toxic. VOCs often have an odour, and some examples include gasoline, alcohol, and the solvents used in paints.

W

Wetland is temporarily or permanently inundated terrestrial systems which border aquatic systems. It also includes the shallow systems such as estuaries, swamps, salt marshes, flood plains and the lagoons and coastal lakes.

Weathering is the physical and chemical breakdown of rocks due to natural process.

Water table is the level of ground water.

Weather is the result of unequal heating of the earth's atmosphere, as a function of terrain, latitude, time-of-year and other secondary factors.

1. pH is defined as
(i) Logarithm of Hydrogen ions
(ii) Negative logarithm of Hydrogen ions
(iii) Hydrogen ion concentration
(iv) OH ion concentration

2. pH of neutral water is
(i) less than 7
(ii) more than 7
(iii) 7.0
(iv) 0.0

3. For pure water at 25°C, the product of H+ and OH– ions is
(i) 10^{-7}
(ii) 10^{-14}
(iii) 10
(iv) 107

4. The acceptable value of pH of potable water is
(i) 7.0 to 8.5
(ii) 6.5 to 9.5
(iii) 6 to 8.5
(iv) 6.5 to 10

5. Acidity of water means
(i) pH of water in acidic range
(ii) pH of water in alkaline range
(iii) base neutralizing capacity of water
(iv) acid neutralizing capacity of water

6. The alum is most effective as a coagulant in the pH range of
(i) 6.5 to 8.5
(ii) 6 to 9.0
(iii) 6.5 to 9.5
(iv) 7.0 to 7.5

7. For the aerobic decomposition of organic matter the pH should not go below
(i) 5.0
(ii) 6.0
(iii) 7.0
(iv) 9.0

8. Following indicator is used for pH determination of water between 4 to 11 pH
(i) Phenolphthalein
(ii) Methyl orange
(iii) Universal Indicator
(iv) Bromthymol Indicator

Correct Answers
1. (ii) 2. (iii) 3. (ii) 4. (i) 5. (iii) 6. (i) 7. (i) 8. (iii)

1. The hardness of water is mainly due to the presence of
(i) Carbonate, bicarbonate, chlorides and sulphates of calcium and magnesium
(ii) Carbonate, bicarbonates of calcium and magnesium
(iii) Chlorides and sulphates of calcium and magnesium
(iv) Nitrate and sulphates of calcium and magnesium
2. The hard water
(i) is not tasty
(ii) is saline water
(iii) consumes more soap for cleaning purposes
(iv) consumes more chlorine as disinfectant
3. The temporary hardness is due to
(i) Carbonate and bicarbonate of calcium and magnesium
(ii) Sulphate of calcium and magnesium
(iii) Chlorides of calcium and magnesium
(iv) Nitrates of calcium and magnesium
4. Water with hardness up to 50 ppm is known as
(i) Hard water
(ii) Soft water
(iii) Moderately hard water
(iv) Moderately soft water
5. The permanent hardness is due to
(i) Sulphates, chlorides and nitrates of calcium and magnesium.
(ii) Carbonate and bicarbonate of calcium and magnesium
(iii) Sulphate and bicarbonates of calcium
(iv) Chlorides and carbonates of magnesium
6. E.D.T.A means
(i) Ethylene diamine tetra acetic acid
(ii) Erichrome diamine tetra acetic acid
(iii) Ethyle dye toluene acid
(iv) Erichrome dye toluene acid
7. The hard water
(i) is corrosive
(ii) forms scales
(iii) is tasteless
(iv) is costly
8. Magnesium hardness with sulphate ions produces
(i) Cancer
(ii) Laxative effect
(iii) Breathing problem
(iv) Tiredness

Correct Answers
1. (i) 2. (iii) 3. (i) 4. (ii) 5. (i) 6. (i) 7. (ii) 8. (ii)

1. Disinfection means
(i) Killing of disease producing bacteria and other microorganisms.
(ii) Killing of all bacteria and other microorganisms.
(iii) Removing infection from water.

2. Bleaching powder is mixed in water for
(i) Making it clean
(ii) Disinfection of water
(iii) Adjusting its pH
(iv) Making it soft water

3. Residual chlorine means
(i) Chlorine required for the disinfection of water normally
(ii) Chlorine required for the disinfection of water in the rainy season
(iii) Chlorine available at the consumer's end.
(iv) Chlorine required as the super chlorination.

4. Potable water is
(i) tasty water
(ii) wholesome water
(iii) Mineral water
(iv) Water free from disease producing elements and bacteria.

5. The amount of residual chlorine in water should be
(i) 0.2 mg per liter
(ii) 2.0 mg/litre
(iii) 2.5 mg/litre
(iv) 4.0 mg/l

6. Residual chlorine is detected in water by
(i) Erchrome black T
(ii) Bleaching powder
(iii) Methyl orange
(iv) Orthotolidine

7. Excessive chlorine in water gives
(i) Bad odour
(ii) Bad taste
(iii) Harmful effect
(iv) All of the above

8. If more chlorine is present in water, the colour produced by orthotolidine is
(i) lighter
(ii) darker
(iii) no difference

Correct Answers
1. (i) 2. (ii) 3. (iii) 4. (iv) 5. (i) 6. (iv) 7. (iv) 8. (ii)

1. Conductivity is
(i) Ability of an aqueous solution to carry current
(ii) Ability of an aqueous solution to dissolve a solid
(iii) Ability of a solution to conduct heat
(iv) Ability of a solution to conduct sound

2. Conductivity depends upon
(i) Presence of ions
(ii) Valence & relative concentration
(iii) Temperature
(iv) All of the above

3. The electric resistance of a conductor is
(i) Inversely proportional to its cross sectional area
(ii) Directly proportional to its length
(iii) Inversely proportional to its cross sectional area and directly proportional to its length
(iv) Directly proportional to its cross sectional area and inversely proportional to its length

4. Specific resistance is the resistance of
(i) A cube of 1 cm.
(ii) One litre of water
(iii) One gallon of water
(iv) None of the above

5. Conductance is
(i) Reciprocal of the resistance
(ii) Square of the resistance
(iii) Cube of the resistance
(iv) None of the above

6. The conductivity of potable water varies from
(i) 50 to 1500 micro mhos/cm
(ii) 150 to 2500 micro mhos/cm
(iii) 500 to 5000 micro mhos/cm
(iv) more than 15000 micro mhos/cm

7. The measurement of conductivity may lead to the estimation of
(i) Total solids
(ii) Total dissolved solids
(iii) Suspended solids
(iv) Colloidal solids

8. Freshly made distilled water has a conductivity of
(i) 2.0 to 2.5 micro mhos/cm
(ii) 0.5 to 2.0 micro mhos/cm
(iii) 2.5 to 3.5 micro mhos/cm
(iv) 3.5 to 4.5 micro mhos/cm

Correct Answers
  1. (i) 2. (iv) 3. (iii) 4. (i) 5. (i) 6. (i) 7. (ii) 8. (ii)

1. **Most common ion in the water is**
(i) Fluoride
(ii) Nitrate
(iii) Chloride
(iv) Sulphate

2. **Chloride gives salty taste to water particularly when present as:-**
(i) Sodium chloride
(ii) Magnesium chloride
(iii) Potassium chloride
(iv) Zinc Chloride

3. **The acceptable limit of chloride in potable water is**
(i) 200 mg/L
(ii) 500mg/L
(iii) 1000 mg/L
(iv) 1500 mg/L

4. **The chloride concentration in sewage is**
(i) More than the water supplied
(ii) Less than the water supplied
(iii) Equal to the water supplied
(iv) None of the above

5. **Chloride consumed by us**
(i) Pass through the faecal matter as it is.
(ii) Gets changed into another form
(iii) Gets disappeared
(v) None of the above

6. **High chloride content in water**
(i) Harms metallic pipes
(ii) Harmful for irrigation
(iii) Harmful to human beings
(iv) All the above

Correct Answers
1. (iii) 2. (i) 3. (i) 4. (i) 5. (i) 6. (iv)

1. Which water generally has high nitrate concentration?
(i) Surface water
(ii) Ground water
(iii) Distilled water
(iv) None of above

2. The ground water has high nitrate concentration because of
(i) Percolating sewage
(ii) Industrial waste
(iii) Chemical fertilizers, leaches etc.
(iv) All of above

3. The high nitrate concentration of water fed to infants causes
(i) Green baby disease
(ii) Blue baby disease
(iii) Anemia
(iv) Cancer

4. The acceptable limit of nitrates in potable water is
(i) 75 mg/L
(ii) 45 mg/L
(iii) 90 mg/L
(iv) 200 mg/L

5. The determination of nitrates is important as to
(i) determine nitrates important from health point of view that is safe against the diseases produced
(ii) assess the self purification capacity of water bodies and the nutrient balance in surface waters and soil.
(iii) find out state of decomposition of organic matter in sewage
(iv) All of above

Correct Answers
    1. (ii) 2. (iv) 3. (ii) 4. (ii) 5. (iv)

1. Fluoride is essential for human beings
(i) To fight against dental caries.
(ii) To fight against fluorosis
(iii) To fight against molten enamel
(iv) To fight against deformed skeleton
2. The desirable concentration of fluorides in potable water is
(i) 3.0 mg/L
(ii) 1.0 to 1.5 mg/l
(iii) 45 mg/l
(iv) 200 mg/l
3. The skeletal fluorosis affects
(i) Bones
(ii) Tendons
(iii) Ligaments
(iv) All of the above
4. The dental fluorosis affects
(i) the root of teeth
(ii) The enamel of the teeth
(iii) The gums
(iv) The jaws

Correct Answers
 1. (i) 2. (ii) 3. (iv) 4. (ii)

1. The concentration of Dissolved Oxygen in water is mainly dependent on
(i) The temperature
(ii) Chloride concentration
(iii) Organic purity of water
(iv) All of the above

2. The minimum Dissolved Oxygen required for aquatic life in general is
(i) 9.2 ppm (ii) 4 ppm
(iii) 8.4 ppm (iv) 12 ppm

3. The treatment of wastewater is mainly done
(i) To satisfy its B.O.D.
(ii) To remove suspended solids
(iii) To remove odour
(iv) To remove colour

4. The allowable limit of BOD of wastewater to be disposed in rivers is
(i) 45 ppm (ii) 30 ppm
(iii) 100 ppm (iv) 300 ppm

5. The Dissolved Oxygen in potable water
(i) imparts freshness (ii) improves taste
(iii) improves smell (iv) none of the above

Correct Answers
    1. (iv) 2. (ii) 3. (i) 4. (ii) 5. (i)

1. Sewage contains
(i) about 99% solids (ii) about 99.9 % solids
(iii) about 99.9 % water (iv) None of above
2. Sewage contains
(i) organic solids (ii) inorganic solid
(iii) all of above (iv) none of above
3. The solids in sewage are
(i) Dissolved solids
(ii) Suspended solids
(iii) Colloidal solids
(iv) All of above
4. The decomposition of sewage depends upon
(i) Temperature (ii) pH
(iii) Absence of toxic matter (iv) All of above
5. The aerobic decomposition leads to gases like
(i) NH3 (ii) H2S
(iii) CO2 (iv) CH4
6. The anaerobic decomposition leads to gases like
(i) CO2 (ii) CH4
(iii) SO2 (iv) NO2
7. The determination of total solids gives an idea about
(i) The foulness of the sewage
(ii) The B.O.D. of the sewage
(iii) The expected load on sedimentation units
(iv) All of above

Correct Answers
  1. (iii) 2. (iii) 3. (iv) 4. (iv) 5. (iii) 6. (ii) 7. (iv)

1. Putrescible solid means
(i) Readily degradable organic matter
(ii) Most offensive matter
(iii) Solids with high BOD
(iv) All of above

2. The solids in sewage may be
(i) Suspended
(ii) Dissolved
(iii) Volatile or non volatile
(iv) All of above

3. The dissolved solids that impose BOD are
(i) Volatile solids
(ii) Non volatile solids
(iii) Inorganic solids
(iv) None of the above

4. The volatile or organic fraction of the solids is observed because
(i) It produces BOD
(ii) It is putrescible
(iii) It consumes the D.O
(iv) All of the above

5. Aerobic treatment is better as
(i) It produces more stable solids
(ii) It produces lesser harmful gases
(iii) It is more hygienic
(iv) All of the above

6. Anaerobic treatment is more desirable
(i) When the concentrated solid organic matter (sludge) is to be digested
(ii) When a cheaper method is sought
(iii) When the end products are to be used, e.g. biogas
(iv) All of the above

Correct Answers
    1. (iv) 2. (iv) 3. (i) 4. (iv) 5. (iv) 6. (iv)

1. Turbidity is caused due to
(i) The suspended particles
(ii) The settleable particles
(iii) The colloidal particles
(iv) All of the above

2. The colloidal fraction consists of particles with dia
(i) More than 1 micron
(ii) 1 millimicron to 1 micron
(iii) 0.15 mm
(iv) 0.2mm

3. The settleable suspended solids with diameter 0.15 to 0.2mm are generally
(i) Inorganic
(ii) Organic
(iii) All of above
(iv) None of above

4. The settle able solids are determined for the design of
(i) Screens
(ii) Grit chambers
(iii) Filters
(iv) Intakes

Correct Answers
1. (iii) 2. (ii) 3. (i) 4. (ii)

www.ingramcontent.com/pod-product-compliance
Lightning Source LLC
Chambersburg PA
CBHW070324190526
45169CB00005B/1740